INSTRUCTOR RESOURCES

Crossing the River with Dogs

Problem Solving for College Students

Ken Johnson
SIERRA COLLEGE

Ted Herr
ROSEVILLE HIGH SCHOOL

Judy Kysh
SAN FRANCISCO STATE UNIVERSITY

Key College Publishing
Innovators in Higher Education

www.keycollege.com

Ken Johnson
Sierra College
5000 Rocklin Road
Rocklin, CA 95677

Ted Herr
Roseville High School
#1 Tiger Way
Roseville, CA 95678

Judy Kysh
San Francisco State University
1600 Holloway Avenue
San Francisco, CA 94132

Key College Publishing was founded in 1999 as a division of Key Curriculum Press® in cooperation with Springer New York, Inc. We publish innovative texts and courseware for the undergraduate curriculum in mathematics and statistics as well as mathematics and statistics education. For more information, visit us at www.keycollege.com.

Key College Publishing
1150 65th Street
Emeryville, CA 94608
(510) 595-7000
info@keycollege.com
www.keycollege.com

Instructor Resources CD-ROM
Key College Publishing guarantees that the CD-ROM that accompanies this book is free of defects in materials and workmanship. A defective disk will be replaced free of charge if returned within 90 days of the purchase date. After 90 days, there is a $10.00 replacement fee.

Development Editor: Allyndreth Cassidy
Production Director: McKinley Williams
Production Project Manager: Beth Masse
Production Coordinator: Ken Wischmeyer
Project Manager: Erin Milnes
Copyeditor: Margaret Moore
Compositor: Thompson Type
Proofreader: Andrea Fox
Art and Design Coordinator: Kavitha Becker
Cover Designer: Todd Bushman
Cover Photo Credit: Superstock, PictureQuest
Printer: Alonzo Printing Co., Inc.

Editorial Director: Richard J. Bonacci
General Manager: Mike Simpson
Publisher: Steven Rasmussen

Printed in the United States of America
10 9 8 7 6 5 4 3 2   08 07 06

ISBN 1-931914-38-9

Contents

Teaching Resources in Detail

Letter to Colleagues

This is not a book. It is a course. Written and oral communication skills are an important emphasis of the course. Students practice these skills in presentations, group interaction, and written solutions.

The course has been carefully written and tested, and we are aware of no other math course like it. Teaching the course requires a high degree of interest on your part. If you have that interest, this will most likely be your favorite course, and a majority of your students will say it was the best course they ever took. Your students will become much better problem solvers and be more willing to take risks as a result of taking this course. Subsequent mathematics courses with these students can be taught very differently because of the problem-solving skills they have learned.

This is not a course for you to show your students how to solve the problems. Although there will be times when you want your students to see you struggle with a problem you have not solved, for the most part you want to let the students solve the problems. Your role is to guide students in developing their mathematics and communication skills. You should plan on solving all of the problems yourself, however, outside of class. We think you will enjoy it.

There will also be a significant change in your students' attitudes. Many will start to enjoy the challenge of this class, and many will come to see themselves as mathematically enabled. The classroom atmosphere you create is an important consideration in this process. A competitive atmosphere is destructive. Students must feel that it's okay to be wrong and it's safe to ask questions—that they have permission to be different and that divergent thinking is not only accepted but encouraged.

You will often need to get out of the students' way when they are solving problems. Students may do things differently than you would or even in ways you think are incorrect. But resist correcting them. Exploration and error are necessary for developing students' communication and problem-solving skills.

Good luck teaching this unique course. We hope you have as much fun with it as we have.

Sincerely,
Ken Johnson, Ted Herr, and Judy Kysh

How to Use These Resources

We begin the *Instructor Resources* with a letter describing our philosophy and some of our expectations for the course. Share the letter with your colleagues. Let them know what the course is about and why you take the approach you do.

The rest of this supplement is organized into four main parts: Teaching Problem Solving, Teaching Resources for Chapters 1 through 17, Sample Midterms and Final Exam Options, and Answers to Problem Sets.

TEACHING PROBLEM SOLVING

In this first section you'll find a discussion on how to organize a problem-solving course, including suggestions for conducting the class on a daily basis, ideas and activities for fostering cooperative group work, and suggestions for assessing your students' problem-solving work.

TEACHING RESOURCES FOR CHAPTERS 1 THROUGH 17

Chapter Overview

Each chapter overview provides a summary of the key points in the chapter, along with amplification of some concepts and comments on didactic strategies for these materials. Here you'll also find notes on problems and Problem Set A from the text, as well as suggestions for good quiz problems.

Text Problems

The problems from the text are reproduced together on pages that can be used to make copies for your students. When it's time for students to work on a text problem, you can ask them to close their books and work from this copy instead so that they won't be tempted to read ahead to the solutions without first trying to solve the problem. Cut the copies in strips if you wish to hand out just one problem at a time.

Questions from the Reading

Questions about the solutions to the problems in the text are included to make sure the students learned what was intended when reading the solution. There are Questions from the Reading for Chapters 1 through 3 only. These are meant primarily as an aid for students who have not developed their reading skills to the point that they can read simple technical text. We have three alternatives for using these questions effectively:

1. Hand out the questions to each group and have each group answer them, either individually or as a group, to turn in or just to discuss.

2. Use some or all of the questions in a whole-class discussion about the reading.

3. Give them as an individual homework assignment.

Problem Set A, Version 2

These problems are similar to the ones from Problem Sets A in the text. Use these problems sparingly if and when you need an extra problem for groups or for a Problem of the Day. They can be written on the board or, if the problems are long, selected ones can be duplicated and distributed. You can use some of these problems for extra practice if your students are having difficulty with the strategy. You may also want to use certain problems for a quiz.

Problem Set B, Versions 2 Through 4

Like the Problem Sets B in the text, these problem sets may require strategies from previous chapters. For this reason, Problem Sets B don't appear until Chapter 3. Because Problem Sets B are the most important component of student evaluation, these additional versions are provided so that you can use different problems in successive semesters. For example, you could use the Problem Sets B in the text for the fall semester, then use Version 2 in the spring or the next time the course is taught. Reserve Versions 3 and 4 for the following year. This way you will have at least two years of fresh problems.

SAMPLE MIDTERMS AND FINAL EXAM OPTIONS

Two sample midterms and their answers are provided. You will note that we do not ask students to solve the problems completely, because we are testing their knowledge of the strategies and their thought processes. We give a group final at the end of each course. Suggested problems for the final exam with answers are also provided. The problems include suggested strategies and their level of difficulty.

ANSWERS TO PROBLEM SETS

All of the answers to problem sets from both the student text and the *Instructor Resources* appear in the Answers to Problem Sets section. Because complete solutions to text problems can be found following the problems in the student text, answers to these problems are not included in the *Instructor Resources*.

INSTRUCTOR RESOURCES CD

Packaged with this text is a CD that contains presentations prepared with Microsoft PowerPoint® presentation software. One text problem from each chapter is discussed with PowerPoint slides. Use these presentations in class as you teach new strategies. Encourage your students to work through the problem before you reveal the next steps in subsequent slides.

You may also modify the slides to make them appropriate for your course.

Teaching Problem Solving

Overview

Philosophy and Goals

Problem solving is a very important part of learning mathematics. Good problem-solving skills don't necessarily come naturally but can be taught. Students need many opportunities to practice problem-solving strategies, and they need to learn how to choose an appropriate strategy to solve a given problem.

As mathematics instructors we learned a lot of strategies on our own and developed them with minimal outside impetus. Some instructors may think that a student loses something by not being left alone to develop problem-solving skills. Our experience tells us that, although some athletes may be naturally good at what they can do, they are better when they're coached. This is also true for problem solving.

We often hear the call to "integrate problem solving into the regular math courses." Our response is that (1) this integration hasn't happened everywhere yet; (2) where it is happening, it isn't enough; and (3) students still see traditional math topics as the point of their math courses, not problem solving—this would be okay if students were already good at problem solving, but before they are capable problem solvers it gives them too much to concentrate on. Should they concentrate on the math content or on the problem-solving strategies? They can't do both from the start. They can do both if they're already good problem solvers.

This course teaches more strategies than come up naturally in even the best problem solving–integrated math courses. For example, an algebra course may employ the strategies of guess-and-check, patterns, working backwards, and subproblems, and it may also touch on unit analysis. Geometry uses the strategies of diagrams, subproblems, and eliminating possibilities. An advanced algebra course may touch on easier related problems, but students may never be exposed to systematic lists, matrix logic, or manipulatives. And they won't fully explore strategies that do arise in other courses.

This course is designed to give students a firm problem-solving foundation. It also teaches them to think and work together, present solutions orally to the whole class, and write up detailed solutions. In other words, it helps to prepare them for life, as well as for future mathematics courses.

Background

We developed and first taught a one-semester problem-solving course at Luther Burbank High School in Sacramento, California, in 1985. Since 1991, Ken Johnson has taught the course at Sierra Community College in Rocklin, California. Originally we designed the course for upper-grade high school and community college students who had completed a year of algebra. The problem-solving course is not easy and requires a certain level of maturity and "stick-to-it-iveness" for students to produce the expected outcomes. This new version of the course has been redesigned specifically for use at the college level, with a prerequisite of intermediate algebra.

Overview of the Course and Textbook

Students using *Crossing the River with Dogs: Problem Solving for College Students* learn a different strategy and its substrategies in each chapter. Each chapter takes about a week, so by the end of the semester, students will learn at least 15 problem-solving strategies and substrategies.

Each strategy, along with some of its practical uses, is described at the beginning of the chapter. Students are then presented with a problem and asked to close the book and try to solve the problem. Complete solutions follow these example problems, using the strategy presented in the chapter. We present most solutions from a student's point of view, using diagrams, charts, and lists, with commentary and explanation provided at every step. These problems and solutions form the body of each chapter. Certain problems reappear in several places in the text to demonstrate multiple approaches to the same problem. It's important that students learn that there are often many ways of solving a problem. Require students to read the text. About four pages each night is reasonable, depending on the natural breaks in the text explanations.

Using the Problem Sets

In addition to the example problems used to model a strategy, two problem sets at the end of most chapters provide students with practice using problem-solving strategies as they learn them.

Problem Set A consists of problems that can be solved by using the strategy presented in the chapter. Students should work on these problems in groups during class or individually at home as the Problem of the Day. (Problems of the Day are described in the Presentations and the Problem of the Day section.)

Problem Set B includes problems that can be solved by any of the strategies presented in that chapter or in any previous chapter. The first Problem Set B appears in Chapter 3. The Problem Sets B are designed to be used as weekly problem sets, with students required to write full explanations of their solution processes. Five problems in Problem Set B may seem like too few. However, five is enough, sometimes more than enough, if you expect thorough write-ups. In problem solving, less is better. Some strategies can be hard to learn, and students need time to develop their skills. The students will probably spend four to eight hours during the week on each Problem Set B. We have found it works best to assign Problem Set B in the week after you finish the chapter, to be due the following week. It's also best if you can give your students some time in class to work in groups on the problem set. As you'll see in the Assessment section that follows, we consider these problem sets to be the most important component of a student's grade.

Important note: In each chapter (starting in Chapter 3) in the text there is a Problem Set B. In this *Instructor Resources,* there are three additional versions of Problem Set B for each chapter. Thus you can alternate Problem Set B versions every semester, to keep students from sharing solutions with previous students and to maintain your own interest by having different problems to work with.

A Typical Chapter

At the beginning of a chapter, depending on the strategy, you might start modeling a new strategy by working one problem on the board. Then have students work several problems from Problem Set A (or the text) in their groups. Some strategies such as diagrams do not require any modeling, and you can start directly with students solving problems in their groups. Then assign a problem from Problem Set A for completion as homework and presentation in the next class. Start the next class with student presentations. Then give students more problems to do in their groups. Starting in Chapter 4, at least part of one class each week should be devoted to having students work in groups on Problem Set B. In the first few weeks of the course, there should be student presentations every day. After Problem Set B starts, you can cut down on Problems of the Day and student presentations. Covering one chapter a week works well. See the First Three Weeks and Beyond section for more details.

The Classroom Atmosphere

The focus throughout the course is on what students are doing. During class, students will spend a good part of their time engaged in problem solving, working in cooperative groups. Much of the first week of the course should be spent building the skills students need to work effectively in groups. (See the Groups section for more information and for ideas for group-building activities.)

Students will spend most of the rest of their class time presenting solutions and discussing those solutions. A safe, supportive atmosphere that allows all students to feel comfortable in presenting their work is essential. Some students may have developed anxiety about presenting in other classes in which problem solving was the "hard part" of the course. And public speaking makes many people anxious. But students who feel supported by one another and who have the opportunity to focus on improving their problem-solving skills without fear of embarrassment or reprisal will develop confidence in their ability and will overcome these anxieties. Read the Presentations and the Problem of the Day section for more on how to establish this safe atmosphere.

Finally, we should mention that some of our students have accused us of using bait-and-switch tactics in this course. You and your students should be warned that the course starts out quite easy but gets harder fast, starting in Chapter 3. Some students complain good-naturedly that they were "lured" into thinking that this was an easy course by the first two chapters before we "switched" it for a tough course. The advantage to starting easy is that students gain confidence and overcome anxiety. These first two or three weeks can set the tone for the course. Once students feel like capable problem solvers, they're ready for the greater problem-solving challenges to come.

Presentations and the Problem of the Day

Every day in the first third of the semester and about half the time in the rest of the semester, give a Problem of the Day (POD) at the end of the period. PODs will generally come from a Problem Set A or from a Problem Set A, Version 2, in the Teaching Resource materials in this book. Students should be told to expect to work on the POD each day for about 10 to 20 minutes, outside of class.

At the beginning of the class period, allow students who are doing presentations time to go to the board to prepare their solutions to the POD. You will find that students start arriving for class early so that they can get a good space on the board. Note that you need a classroom that has lots of boards, preferably on every wall. During the first few weeks, have students write their names and state their names when they start presenting so that the class can learn everyone's name. If your board space can't accommodate all the people who want to present solutions every day, you can have students write out their solutions on blank overhead transparencies with overhead projector pens, then have them present with the overhead. Whatever solution you find, don't waste time having students work on a single board one at a time, erasing between presentations.

During this time you may want to walk around the room and check the POD for everyone else who did it. Then have everyone (including you) sit down and call up the students one at a time to present their solutions. Each presentation should last between 30 and 60 seconds. Insist that each person give a clear explanation of how she or he solved the problem, not just the answer. Even if a student solved a problem in essentially the same way as a previous presenter, he or she needs to explain the solution as if no one else had presented it. When the explanation is finished, allow the presenter time to field questions from the class. Beyond asking whether there are questions for the presenter, and saying "thank you" at the end, don't make any further comment, whether positive, negative, or correcting. The total presentations will probably take between 15 and 25 minutes.

It is important that wrong answers are presented too. Don't allow your students to avoid presenting incorrect solutions. Often, productive debates center around what the right answer is. These debates would never occur if you allowed only right answers to be presented.

You must create a safe, supportive atmosphere so that all students feel comfortable presenting. Questions should be directed toward what the presenter did. Don't allow students to criticize each other or ask questions like "Why didn't you do it this way?"

Also, do not get into a dialogue with each student as they are presenting. Let her or him talk to the class, not to you. This will build students' confidence in public speaking and take the focus away from you. At first you will need to remind the presenters that they are to address the class, not you.

After all the presentations have been made, you may want to briefly summarize them, but do not give your solution. Even if the students could gain some valuable insight, the short-term opportunity is not worth the long-term loss. The students must not think that you are going to rescue them with a better way.

Even if no student challenges incorrect work, you must let it slide. If they believe that the instructor will take care of cleaning up their mistakes, they won't be concerned about doing it themselves. You must not give your method for solving a problem, even if the students didn't come close to the most effective manner. To encourage divergent thinking, you must value it. That includes not setting the stage for your thinking to compete with that of the students. This approach will build confidence in your students, and they will start to look to themselves as problem solvers rather than to you as the ultimate word on everything.

It's important to mention the variety of solutions and to summarize the similarities and differences among the different approaches. Students will ask you if they are right or who is right if there are multiple answers. Resist the urge to tell them. If they are unsure who is right, let them discuss it in groups and maybe carry it over to the next day. Avoid telling them that their answers are correct. We have never given a problem that everyone missed. They will eventually resolve the differences.

Two of the most interesting problems to give as PODs are Soccer Game on page 7 and Storage Sheds on page 39 of the text. Soccer Game has many correct answers, depending on the interpretation of what constitutes a "five." We usually give this problem as the very first POD, because many different answers will be presented. Accept all of the answers without comment. Storage Sheds also should generate some interesting discussion of whether, for example, a 10-by-8 shed is the same as an 8-by-10 shed.

Evaluating the Problem of the Day

The Problem of the Day generates two different grades: one for doing the problem, the other for presenting. We count each of these grades for 5–15% of the total grade in the course.

To give the students credit for doing the problem, we walk around the room during the first few minutes of class as the presenters are writing on the board and simply check off the names of those students who did the problem. We allow students to make up these problems after absences, because it is important for everyone to do every problem assigned. The solution does not have to be correct to get credit. If a student doesn't make a reasonable effort to solve the problem, then he or she will need to rework it and turn it in later.

We grade presentations separately, with presentation points, and we require a certain number of presentations for various grades. The standard we tell our students at the beginning of the quarter or semester is that they have to do one presentation every two weeks to get a C in presentations. If they do more, they get a higher grade.

Some students will want to present all the time, and they tend to monopolize the available board space. Other students are easily intimidated and won't go to the board without some prodding from you. Remind those students of their obligation to do one presentation every two weeks. The first few presentations are the toughest. After that, most students are willing to go to the board at any time. Creating the safe atmosphere described earlier is vital to many of your students' willingness to present. As you walk around the room checking for work, note the students who have not done many presentations. If their solutions look interesting (and correct), encourage them to put their solutions on the board. This will give shy students a chance to present their best work and will give them confidence in presenting.

Groups

This class is built around groups. The instructor's role is often just to listen and to stay out of the way. If you have rarely taught using groups, this will be a major change in approach. This class also emphasizes process more than answer—the journey is more interesting than the final destination. Very few problems can be approached in only one way. Traditionally, an instructor often directs mathematics classes and students merely follow the instructor's examples. *Problem solving develops better when the students have to determine their own direction through a problem.*

If you haven't had students work in groups before, allowing them to work together may seem strange. You may wonder what individual students can actually do by themselves. You should feel more comfortable about that issue after you have used groups for a while. For this course, we recommend that you encourage your students to work together. You may wish to give an occasional quiz of one or two problems for each student to do individually. See the Quizzes and Midterm Exam section for more about quizzes.

What are the benefits of using groups?

Students often feel intimidated about asking the instructor questions when they don't understand something. A small group of four students is much less intimidating, and most students will be more willing to admit that they are confused in a small group of their peers.

Many problems in this course are very difficult, and some students could not solve them completely alone. However, when four minds get together to solve one problem, ideas build on other ideas and solutions are attained more easily. There is tremendous benefit to observing someone else's thinking process. We sometimes convince ourselves that our students achieve this insight into thinking by watching us solve problems. The problem with that belief is that we already know how to solve the problem and our students know this. When they observe a peer's thinking processes, they know that their peer is struggling to solve the problem just as they are. Groups also allow for more on-task behavior than standard seating. When a class is motivated and working, the students are close to 100% on task. The learning in groups is active learning, not passive learning. And students are learning valuable life skills as well as job skills. People who can work well with others have good tools for success in their careers.

Research into cooperative group work indicates that students learn more, retain what they learn better, and have a better attitude about school and about themselves. You may also find that students who are accustomed to working in groups are more likely to work together outside of class.

How should I arrange my classroom?

The physical arrangement of the classroom is very important. A sample class arrangement is shown below. Suppose your classroom has 6 rows of desks, with 6 desks in each row. These 36 desks can easily be organized into 9 groups of 4. Students turn their desks sideways and face each other.

If you walk around the center group, you will walk next to every group and be able to listen in on every group. Move through the room in different ways, though, to be sure that you are not ignoring anyone. Do not just stay at your desk or in one part of the room.

Suppose your class has eight rows of five desks. This causes more of a problem. If you have 32 students or less, just make eight groups (or less). If you have 36 students, you may want to add one student to each of the back groups and have four groups of four and four groups of five. You may also want to try to create a ninth group in whatever section of the room has the most space.

Be creative if your classroom situation is less than desirable, but don't use a bad classroom situation as an excuse to not try groups. Virtually any classroom can achieve as good or better space and traffic efficiency with desks arranged in groups, whereas the class will suffer if students are always sitting in rows. With practice, the conversion from rows to groups or from groups to rows can be done in less than a minute.

Desks within groups must touch. You should not be able to walk between members of a group. If students are not sitting close enough together, then ask them to move, or move their desks yourself. Each group member must be able to communicate with all others in the group. If the desks are close together, then you should have enough room to walk between different groups. Set up a walking path around the room that brings you next to every group in one circuit around the room.

How should I select students for the groups?

We select students for groups randomly by one of the following methods:

1. From a deck of cards, take enough cards to represent the number of students in the class. For example, if you have 24 students, you would use the aces, twos, threes, fours, fives, and sixes. Then shuffle the cards and give a card to each student. All the aces are in one group, the twos are in another, and so on.

2. Have each student fill out a three-by-five card with his or her name, address, phone number, and any other relevant information. Keep these all quarter or semester. Use these cards to change groups. Shuffle the cards and deal them out, four per group. Then call off the names that form each group.

3. Go around the room and count off, giving each person a number according to how many groups you want. If you have 32 students and you want eight groups, count off 1 to 8, then start again at 1.

What if the number of students in my class is not a multiple of four?

In general, it is better to form groups of five rather than groups of three. A group of three loses a lot of communication avenues and is decimated if one person is absent.

What do I do about absences?

If one person is absent, leaving a group of three, it is probably okay for one day. If two people are absent, you will probably want to move the two remaining people to another group or put them in two different groups. You may also want to do this if there are several groups of three one day.

How often should I change groups?

Change groups once a week for the first few weeks, then every two weeks or so after that. Changing groups early and often allows students to learn the names of their classmates sooner. It also lets them experience multiple viewpoints and gives them the idea that everyone in the class can contribute something to the group process. If you leave students in the same groups too long at the beginning of the semester, they may get too comfortable and resist changing groups, fearing that a new group won't be as productive.

What is the instructor's role?

The instructor's main role is to get out of the way. You are no longer "the sage on the stage." Your new role is that of "the guide on the side." Avoid interrupting a group to give them your opinion or direction. Do keep them on task, however. You need to guide, not direct. Steer them in a better direction if they get way off track and if you don't think they will get back on. Many times, students will get back on track by themselves, so let them try. The kind of communication and thinking that occurs in these situations is invaluable. It is difficult to know when to stay out of your students' way and when to interrupt. If you can't tell what to do, it's better to leave them alone, but keep circulating and listening. Ask questions, avoid giving answers. Don't tell students whether or not they are right—that tends to shut off their thinking. To become better thinkers, students need to evaluate situations (problems and answers) for themselves. If they find that their instructor will always be there to provide a crutch, they will learn to limp.

Group-Building Activities

At the beginning of the course, we use activities that encourage students to work together. Two very good activities are the Color Square Game and the Digit-Place Game, described in this section. Some other good activities follow. The first four activities help students get to know one another and learn one another's names.

Interview: When the students first move into new groups, have them pair up and interview their partners. Allow about one to two minutes, then have them switch. Then ask each person to introduce her or his partner to the rest of the group.

Jigsaw Puzzle: Here's another great first- or second-day group-building activity: Bring enough 100-piece jigsaw puzzles so that each group can have one. You're likely to need seven to ten puzzles for a class. Take the puzzles out of their boxes and put them in plastic bags that seal with a "zipper." Do not include the picture that came on the box. Give a puzzle to each group to put together. This is a fun activity that takes about 20 minutes. One-hundred-piece puzzles are perfect because they are generally easy for a group of three or four students to put together without having the picture as a reference.

Fascinating Facts: On the first day of class, have each student write down a fascinating fact about himself or herself. It can be about anything: hobbies, interests, sports, places they have been, where they work, and so on. Insist that the facts be publishable. Put a fact in for yourself too. Then type up the list of facts and distribute one to each student. Two or three times during the next few days, let the students have some time to roam around the room and discover other people's facts. Don't let them just tell one another their facts. Treat this activity as a game and model good questions. For example, "Does your fact have to do with travel? Does it have to do with pets? Does it have to do with hobbies or interests?"

Name Test: After three or four weeks, give your class a low-key name test. Have everyone stand in a large circle. Then pick someone to go first. She says her name. Then the next person says his name and the name of the person who came before. The third person has 3 names to say, the fourth 4, and so on. After the list gets up to 15 or so names, cut it back a little by telling the next person she has to go back only as far as John, who is maybe standing halfway back. After finishing the entire circle, start back with the first people again so that the names from the last part of the circle get mentioned more times.

Teamwork: Have the students individually try to make as many words as possible out of the letters of the word *teamwork*. After a few minutes, allow them to work together in groups.

Making 24 (or some other number): Have the students individually try to make 24 by using one or more mathematical operations or symbols or functions. Then have them work on this in their groups.

The 1776 Game (or some other important year you'd like to use): Have students try to make the numbers from 1 to 100 (or some other smaller range) using the digits 1, 7, 7, and 6, in that order. You can eliminate the order restriction to make the game easier. Years in the twentieth century tend to generate interesting combinations. Note that years in the twenty-first century won't work well with this game for a while. Students can play this game with their birthdates, important years in history, and the like. Here are some examples for the year 1776:

$$1 + 7 \div 7 + 6 = 8$$

$$1 + 7 - 7 + 6 = 7$$

$$1^{776} = 1$$

$$1 + (7 + 7)6 = 85$$

$$(1 + 77) \div 6 = 13$$

Maintain a wall chart with the number and the solution. As students discover solutions for the numbers from 1 to 100, they can write them on the chart.

The Color Square Game (a.k.a. Rainbow Logic)

The Color Square Game is one of the best logic activities that we know. It never fails to excite the students. Play it as a whole class, in small groups, or both ways. The Color Square Game is an easy game to understand, but very challenging to play. The reasoning that students do is quite sophisticated, and they also get practice in explaining their reasoning to others.

The rules of the game are simple. Use a three-by-three square like the one shown at right. Color the squares in groups of three with three colors: red, blue, and green. Squares of the same color must be contiguous (joined on a side). Do not show students your colored squares, but draw a blank three-by-three square on the board or overhead projector. Tell students it is up to them to logically figure out what each square is colored, using as few clues as possible.

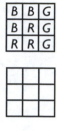

Students begin asking for the colors in particular rows or columns. For example, a student might ask, "What is in row 1?" You might reply, "Row 1 contains two blues and one green" and write "2B, 1G" next to row 1. Do not divulge the order of the two blues and one green. It is best if you always reply by listing the colors in alphabetical order. Ask if anyone knows the color of a particular square. Tell students not to guess but rather to figure things out logically. Always ask why a student believes a square is the color he's chosen. If the student can prove that a square must be a particular color, then fill in that square. We like to let the students explain their reasoning to their group. After all squares that can be filled in are filled in, allow the students to ask another question. Continue the game in this fashion until the students deduce the colors of all nine squares.

Play the three-by-three game a few times. Then graduate to the four-by-four game, for which there are 16 squares: 4 of each of 4 colors. This is a much better game. (We don't recommend the five-by-five game.) The four-by-four game can be played in a number of ways. It is a good idea to discuss with the students how the colored squares can be arranged. In fact, a good exercise for the students is to ask them to find all the ways that four squares of the same color can be put together so that they are touching along at least one entire side.

Manipulatives will make the game and the associated thinking accessible to more students quickly. Give each group four squares of each color, and have them draw a large four-by-four square on a piece of paper. The group manipulates the colored squares to try out different arrangements. One person in the group writes down the group's conclusions on a blank four-by-four square. It is not necessary to use manipulatives to play this game. You may want to try it a few times with manipulatives and a few times without to see what you and your students like better.

Make the solution for yourself, then introduce the game by asking for a student to request a column or row. Post that clue and let students discuss possible conclusions in their groups first. Then have several students explain their group's reasoning to the class. Keep asking if anyone can deduce more squares than are shown. When no more squares can be deduced, let someone ask for another clue. Depending on the clue, this process may be quick or it may take a few minutes. The following is a sample game.

There are four squares of each color: red, yellow, blue, and green. Each color is abbreviated by using its first letter. In this game, the first clue asked for is the fourth column. This column is composed of one yellow and three greens.

Y, 3G

The colors are not necessarily given in order or out of order. This clue in particular would generate a fair amount of discussion as to where the yellow goes. At some point, a student should be able to deduce that the yellow must go at either the top or the bottom of the row, because if it were placed in the second or third row there would not be enough greens to provide a contiguous figure.

Remember that the students are required to *prove* which squares are a certain color, not just to provide possibilities. As the students explore the possibilities on this square, they will realize that whenever three squares are available and two are a given color, the middle square must be that color. For example, here are two possibilities for green being in the middle square:

Y, 3G

The next clue requested is the second row. The students should be able to conclude that the middle square of the three squares remaining in the row is blue, because otherwise the blues could not connect.

				Y, 3G
R, G, 2B		B		G
				G

At this point the next clue the students request is the fourth row. Using this in conjunction with the fourth column, it is quickly evident that the last three squares of the fourth row are all yellow and that the first one must be red.

				Y, 3G
				G
R, G, 2B		B		G
				G
R, 3Y	R	Y	Y	Y

That leads to identifying the top square in the fourth column as green. As the students examine this setup, they realize that the red in the second row must appear in the first column, because it cannot be in the third column and still connect to the red in the fourth row. That also leads to the conclusion that the square in between must be red in order for the reds to connect. The other two squares in the second row must be blue.

				Y, 3G
				G
R, G, 2B	R	B	B	G
	R			G
R, 3Y	R	Y	Y	Y

At this point the game becomes more difficult, but actually one more square can be proven: the second square in the first row. The proof is indirect: By assuming in succession that the square in question is more than two red, or yellow, or green, the students should find that it cannot be any of those because in each case those colors cannot connect to that square; thus, the only possible remaining color is blue.

				Y, 3G
		B		G
R, G, 2B	R	B	B	G
	R			G
R, 3Y	R	Y	Y	Y

Choosing the next clue at this point has to be done carefully. Row 3 must contain the fourth yellow and could contain the fourth blue, green, or red. If it contains the missing blue and yellow, then asking for that clue will be inconclusive as to their exact location. The second column would also be inconclusive if it shows two blues and two yellows, because either of the two squares in column 3 could be either blue or green. Asking for the clue for row 1 would tell the students where the final red goes, but if the red is in that row, they will have the same problem with where the blue and the yellow are placed. So at this point, there is no guarantee that the next guess will end the game. Let each group decide on their own what to ask next and walk around the room giving each group their own clue.

The next clue request in this sample game happens to be the second column. The game basically ends now by filling in each of the colors in succession: the missing yellow, the missing green, and the missing blue.

		2B, Y, R		Y, 3G
		B		G
R, G, 2B	R	B	B	G
	R	R		G
R, 3Y	R	Y	Y	Y

The completed game is shown at the top of the next page.

	2B, Y, R		Y, 3G
B	B	G	G
R, G, 2B → R	B	B	G
R	R	Y	G
R, 3Y → R	Y	Y	Y

After you've played the game several times with the whole class, you may want to let students play the game in their groups. One person makes up the solution, and the other three solve it. Note that students tend to be too willing to make false conclusions. You need to set the tone by having each person who makes an assertion prove it, step-by-step. In another session, have students play the game in pairs to ensure that each student is an active participant. Suggest to students that they play the game at home with friends and family.

You can develop the game a couple of steps further. Write the following examples on the board or on an overhead projector with the instructions to "fill in as many squares as you can prove." The problem on the left has two squares that cannot be proved, and the one on the right can be entirely proved.

	G, Y, 2R					G, 3R		3Y, B
B, 3Y								
				2Y, G, R				
2R, 2B				B, 2R, Y				

Below is another variation of the game. The object of the game is to fill in as many squares as you can prove are a given color.

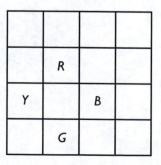

The Color Square Game develops step-by-step reasoning, small-group interaction, the concept of proof by contradiction (indirect proof), and an atmosphere of questioning and substantiating. Solving one of the previously started games as a writing assignment is an excellent introduction to proof.

See *Make It Simpler* by Carol Meyer and Tom Sallee for more information about the Color Square Game.

The Digit-Place Game

Another good group-building game is the Digit-Place Game. The game exists in several forms, including Fermi-Pico-Bagel, a letter version called Jotto, and a commercial version called MASTERMIND. There is now a TV game-show version called Lingo. The version we present here appears in a number of sources, including *Make It Simpler*.

The game is relatively simple to learn. One person chooses a three-digit number with each digit different. The others make guesses, and the person who knows the number provides hints. The hints relate how many of the digits are correct and how many digits are in the correct place. The hints do not include *which* digits are correct or in the right place.

As an example, let's use the number 487 as the correct number. The person running the game chooses this number and does not reveal it to anyone. In this chart, the *D* stands for "digit" and the *P* stands for "place."

Guess	D	P
541	1	0

The 4 is a correct digit, but it is not in the correct place. The person guessing has no way of knowing which digit is the correct one.

Guess	D	P
541	1	0
390	0	0

Now there are no correct digits, so none can be in the correct place.

Guess	D	P
541	1	0
390	0	0
267	1	1

The 7 is both a correct digit and in the correct place, hence a 1 is marked in both columns. The people playing do not know for which digit the tallies were made.

Guess	D	P
541	1	0
390	0	0
267	1	1
880	1	1

The 8 in the tens place is both correct and in the right place.

At this point, all digits have appeared at least once. The second guess was probably the most helpful, because it indicated that all of those digits are not in the right answer. Since 0 is one of those digits, when it reappears in the fourth guess we know it is not the one that supplied the mark in the digit and place columns. Therefore, the 8 is definitely one of the numbers, and it is also in the first or second position.

In the first guess, one of the digits from 541 is correct. The tallies on the third guess indicate that one of the digits of 267 is also correct. We will continue guessing, though we will not necessarily be demonstrating the most efficient method for reaching the number. That is left for the reader to discover!

Guess	D	P
541	1	0
390	0	0
267	1	1
880	1	1
856	1	0

None in the right place helps—the 8 must belong in the center position. Neither the 5 nor the 6 is correct as the 8 is known to be the digit that produced the tally for a correct digit.

Guess	D	P
541	1	0
390	0	0
267	1	1
880	1	1
856	1	0
481	2	2

The 8 produces one tally; the other must be produced by the 4 as we know that the 1 is not in the correct position from the clues for the first guess. Thus, the number is known to be 48__ at this point.

Guess	D	P
541	1	0
390	0	0
267	1	1
880	1	1
856	1	0
481	2	2
487	3	3

By using the third guess and clues, you can see that the 7 is the correct digit since the hint also indicates that the correct digit on this problem is in the correct position.

This game involves a myriad of guessing strategies that will allow the students to get the number within a certain number of guesses. You can also modify the game to allow numbers that have two identical digits or even three identical digits (such as 232 and 777).

Play this game with your class a few times. After each guess, allow students a few minutes to discuss possible conclusions in their groups. Ask a few students to summarize conclusions reached so far. Have another student make a guess and repeat the process. After playing as a whole class for a while, have students play this in their groups. One person runs each game by determining the number and providing the digit-place clues. Have group members take turns running the game. The students should next play it in pairs to ensure full participation.

The game can also take the form of paper-and-pencil exercises for which you provide a set number of guesses and clues. The students must look through those clues to find the most useful ones to solve the problem. Or give the game as a homework assignment: "Play this game with somebody in your household ten times." Playing without a time limit may result in your students playing for a long time and developing strategies at home.

The game can also be played so that students make ten guesses without getting any clues. The game director then provides clues for all ten guesses. The players must declare the correct number on the eleventh guess. This helps the students focus on guessing strategies instead of on the immediate task of determining the number.

Actively encourage the students to look for guessing strategies and have the students verbalize these as they develop them. Verbalizing the strategy is important in helping students develop precision in their mathematical communication.

One effective strategy is to write down the digits from 1 to 10. As you confirm that one number is valid, circle it. Cross off any number that you can eliminate from consideration.

Here are a couple more digit-place examples. Good luck finding the answers.

Guess	D	P	Guess	D	P
158	1	0	837	1	0
269	2	0	295	1	0
370	0	0	160	1	0
482	0	0	416	0	0
591	2	0	905	1	1
725	0	0	874	1	0
803	0	0	529	1	0
452	0	0	604	1	1
398	1	0	321	1	0
965	2	1	537	1	0

Journals

Journals are a great way to get a dialogue going with your students, especially when you write comments back to the students about what they wrote. The following paragraph comes from our syllabus for the course, which is handed out on the first day.

> Every one to two weeks, I will give you a journal topic to write about. You should write whatever you want, giving each topic the respect it deserves and giving some thought to your comments. Please be honest, and don't just tell me what you think I want to hear. I will write back with equally honest commentary. The hope here is that we get a new kind of dialogue going. You will not be penalized in any way for constructive criticism about the class. Please number the journal topics so I can keep track. Journals will be checked off (not graded) and you will get credit for doing them. You will have approximately one week to complete each journal entry and turn it in, either on paper or via email. Due dates for each journal topic will be announced in class and *late journal entries for that topic will not be accepted beyond one week after those dates.* If you have a personal objection to any of the topics, then write about something else that week.

Following are examples of journal topics we have tried that have worked well. We announce these to the class and give students about a week to write about them. Journal topics should always be open-ended, and students should be encouraged to take topics in any direction they choose. The length of a journal entry is up to the student. The order of the topics below is not important, although you will probably want to assign topics 1, 2, and 3 at the beginning of the course.

1. Why are you taking this course? What do you hope to learn?

2. Every student chooses a grade for himself or herself in every course and behaves according to the grade he or she chose. What grade did you choose for yourself in this course, and how are you going to get it?

3. Call or email at least three other students in the course and talk for a while. Write about the experience.

4. Write your reactions to one of the articles or quotations your instructor has provided. [Find interesting articles or quotes, and hand them out for students to read and react to. Some examples appear on the next page.]

5. Discuss how you feel about working in groups. Has it helped you to learn? Have you made some new friends?

6. Keep a time journal for one week. Record all the time you spent during the week for everything: classes, studying, socializing, working, and so on. Then write about what you learned about the way you spend your time.

7. Reflect on your experience in this course so far. [This journal topic is suggested in Reflection, page 130 in Problem Set A for Chapter 5 of the text.]

8. Reflect on your own effort in this course so far. How are you doing toward achieving the grade you chose for yourself at the beginning of the course?

9. Discuss the textbook. Have you been reading it? Has it helped?

10. How do you like writing journals?

11. How have you liked giving and listening to presentations? What did they add to your educational experience?

12. Look back over the whole course and describe what worked and what didn't. How did you like the course?

Following are examples of the kinds of quotations you might hand out for topic 4.

- A former Sierra College student wrote: "Attitude is a small thing that makes a big difference."

- The following statement was found on the back of a student body card: "It's your attitude, not your aptitude, which determines your altitude in life."

- In the movie *The Empire Strikes Back,* when Luke says he'll "try" to raise his X-wing out of the swamp, Yoda says, "Do or do not; there is no *try.*"

- Benjamin Franklin said, "People who are good at making excuses are rarely good at anything else."

- Henry Ford said, "If you think you can or if you think you can't, you are right."

Assessing Problem Solving

Determining what students can do mathematically means looking beyond what students can compute to how well students reason and communicate. This change has been driven in part by the availability of calculators and computers. At this point, those two tools can compute faster and more accurately than a human. But humans are better at reasoning. Therefore, computation should no longer be the major focus of teaching mathematics. The focus needs to shift toward teaching mathematical reasoning—and this book is designed to help you do just that. An integral part of teaching mathematical reasoning is a view of assessment as part of course instruction.

One of the goals of this course is to develop students' communication skills, both in general and in particular, when dealing with mathematical topics. Another goal is to develop their ability to solve problems. We designed the solution write-ups and the method of assessing the write-ups to further both of these goals.

A solution write-up is like a take-home test. The student has access to resources, including classmates and class notes. The final product should be excellent. Your feedback on students' write-ups is necessary to produce the desired changes in attitude and development of skills.

Overall Grading for the Course

There are as many ways of assigning grades as there are mathematics instructors. However, in a unique course such as this, it may be helpful to mention how we assign grades. We split the course's evaluation components into three main parts:

Individual accountability: quizzes and the midterm

Individual/group accountability: problem sets and the final exam

Participation: Problems of the Day, oral presentations, journals, and class participation

Methods for grading Problem Set B are described in detail later in this section. Problems of the Day, their presentations, and journals were described earlier. Instructions for a final exam, a method for grading the final, sample final exam problems, and two sample midterms are featured at the end of the Teaching Resources section.

We suggest that you don't give tests (except possibly a midterm) because of the time limit for solving a number of problems. An occasional one- or two-problem quiz is a good idea. Quizzes and the midterm are described later in this section. You may also want to give and grade a few complex problems as long-term projects. Again, there are many ways of grading.

Grading Problem Set B

As mentioned, starting in Chapter 3, each chapter ends with a Problem Set B. *Note:* There is one version of Problem Set B in the text, and there are three more versions of Problem Set B in this book. The students should have one week to do the five problems, write them up, and turn them in. We suggest using one of the grading rubrics provided in the following pages the first time you teach the course.

Problem solving is difficult to test in a class period. The issue of testing is constrained by the time allotted for the class and by the fact that in the real world those who solve problems in their work are not usually required to do it on their own. Some of the best problem solving takes place when you consult with the right people. Part of problem solving is distillation time—you often need to let problems simmer in the back of your mind for a while before you can solve them. For these reasons, we minimize test giving. Instead, we rely on the problem sets as "take-home tests."

As the quarter or semester wears on, students may start to feel more pressure about getting Problem Sets B in on time. To alleviate some of this pressure in the second part of the course, we've experimented with allowing some Problem Sets B to be turned in by groups of two or three students. Our policy is to require students to turn in their first six Problem Sets B by themselves. So, if a student hasn't turned in one or more problem sets, he has to wait until he turns in six of his own before he qualifies. After that, they can turn in one problem set in a group for each one they turn in by themselves. That is, after they've turned in seven by themselves, they can do one in a group, then one by themselves, then one in a group, and so on. They can also save up several group credits by doing several problem sets in a row by themselves (again, after the first six) and then doing several in a group. They each still need to solve all the problems, but they save time on writing them up.

We require the write-ups to be fairly divided among the group (as fairly as you can divide 5 by 2 or 3). The group also has to document with a Venn diagram how they allocated their time. A Venn diagram for a group of three would show three intersecting circles representing the three people. In regions that don't overlap, they write the time (working alone) spent by the individuals represented by those regions. Time spent working in pairs or with all three group members is written in the appropriate intersecting regions. Every student in a group receives the same score for a group problem set. The students appreciate the flexibility offered by the group option, and it does save them time.

We have provided two versions of a grading rubric. They are our adaptations of the grading system described by Randall Charles in his series of books, *Problem-Solving Experiences in Mathematics* (Dale Seymour, 1994). The system assigns point values to different parts of the problem-solving process. Using this rubric helps make the assessment more objective; it also helps you give valuable feedback to students in a way that is directed toward specific course goals and expectations. We go over the rubric with our students, copying it and handing it out to them.

Grade the papers with an ordered quintuple of numbers (2, 2, 2, 2, 2) for a total of 10 points per problem.

A. Understanding the problem

0—Completely misinterprets the problem.

1—Misinterprets part of the problem.

2—Shows complete understanding of the problem.

B. Choosing a solution strategy (include the name of the strategy)

0—Does not give evidence of using a strategy or uses a totally inappropriate strategy.

1—Chooses a strategy that could possibly lead to a correct solution, or chooses a strategy that will get him or her partway through the problem but fails to change strategies when appropriate. Also, uses a correct strategy but fails to state the name of the strategy.

2—Chooses a correct strategy that could lead to a correct solution if used without error.

C. Implementing the strategy

0—Makes no attempt to solve, uses a totally inappropriate strategy, or uses a correct strategy totally incorrectly.

1—Implements a partly correct strategy based on interpreting part of the problem correctly, or chooses a correct strategy and implements it poorly.

2—Implements a correct strategy with minor errors or no errors.

D. Getting the answer

0—Gets no answer, fails to state the answer, or gets a wrong answer based on an inappropriate solution strategy.

1—Makes copying error or computational error, gets partial answer to a problem with multiple answers, or labels answer incorrectly.

2—Gets correct answer, states it, and labels it properly.

E. Giving an explanation

0—Makes no explanation or incoherent explanation.

1—Gives an incomplete explanation, or the explanation is hard to follow.

2—Gives a clear, coherent, complete explanation.

A. Understanding the problem: The first and second points are for demonstrating an understanding of the problem. Understanding (or misunderstanding) is generally conveyed through the student's work, including the written explanation of the solution. Ideally the student is placing before you a comprehensive map of how he or she solved the problem. This map should also show you exactly where the student made turns, both right and wrong.

B. Choosing a solution strategy: The next pair of points is for choosing a reasonable strategy. The student must also give the correct name of the strategy. A student would get a 1 rather than a 2 if he or she picked a strategy that is cumbersome but that could lead to a correct solution. You have to be careful here because a student may use a strategy that you had not thought of, even though it is a perfectly reasonable one. As you will see as you go through the book, there are many ways to solve many of the problems, and for some problems there may be more than one equally good choice of strategy. Some problems require multiple strategies to solve. If the students name one of the strategies, don't make them tell you all of the strategies they used; it is enough if they name one they used, preferably the major one.

C. Implementing the strategy: The next pair of points is for implementing a reasonable strategy. Here the focus is on how well the students are using whatever strategy they picked. If a student chooses an appropriate strategy but doesn't apply it sensibly, he or she would get zero points here. For example, a systematic list that is completely unsystematic receives zero points; if it is somewhat systematic but has some entries out of sequence according to the system, give one point here.

D. Getting the answer: The seventh and eighth points are for stating the correct answer explicitly. The point is partly to make sure that the student gets the right answer. That is important. It is also important that the student state the correct answer clearly, because sometimes the question asked may be different from the one the student answers. An answer that appears somewhere in the work but is not explicitly stated lacks the authority of a definitive answer. Give zero points here for a correct answer that is never explicitly stated. With practice and guidance, students will quickly learn to state their answers explicitly. A correct answer missing appropriate units should receive one point.

E. Giving an explanation: The final pair of points is for the explanation. The students' explanations should be well-thought-out and well-communicated "maps" of the reasoning they used to solve the problem. Not every detail is needed—students do not need to provide a stream-of-consciousness novel. Rather, the explanation is an organized, concise retelling of the thoughts, assumptions, and understandings students developed while they solved the problem. This exercise in writing out mathematical reasoning will be helpful practice for all kinds of writing students will be asked to do both in school and in their careers. Because explanations for matrix logic problems tend to be very long, we did not include many of these problems in Problem Sets B.

Grade the papers with an ordered quintuple of numbers (2, 2, 2, 2, 2) for a total of 10 points per problem.

A. Trying the problem

0—Does not try the problem at all.

1—Tries a little.

2—Gives a good attempt.

B. Understanding the problem

0—Completely misinterprets the problem.

1—Misinterprets part of the problem.

2—Shows complete understanding of the problem.

C. Choosing and implementing a solution strategy

0—Makes no attempt to solve or uses a totally inappropriate strategy.

1—Chooses a partly correct strategy based on interpreting part of the problem correctly, or chooses a correct strategy and implements it poorly.

2—Chooses a correct strategy that could lead to a correct solution if used without error, and implements it with minor errors or no errors.

D. Getting the answer

0—Gets no answer, fails to state the answer, or gets a wrong answer based on an inappropriate solution strategy.

1—Makes copying error or computational error, or gets partial answer to a problem with multiple answers, or labels answer incorrectly.

2—Gets correct answer, states it, and labels it properly.

E. Giving an explanation

0—Makes no explanation or incoherent explanation.

1—Gives an incomplete explanation, or the explanation is hard to follow.

2—Gives a clear, coherent, complete explanation.

Explanation of Version Two

If you have a weak class, consider using Version Two rather than Version One. Version Two is also good to use with quizzes for which you tell students the strategy to use to solve the problem. As you can see, Version Two is basically the same as Version One, but "trying" points have been added and the strategy and implementation points have been combined into one category. Comments on the changes follow.

A. Trying the problem: The first two points are for trying. This is one case where the student can ask, "Doesn't trying count?" and you can answer, "Yes, for two points." The idea is to encourage students to put enough down on paper so that they can get the full two points for "trying."

C. Choosing and implementing a solution strategy: The two strategy categories have been combined into one. Give both points if the student chooses a good strategy and uses it well. Give one point if the choice of strategy is poor but workable, or if the choice is good but the execution is slightly flawed. With a quiz for which you told students which strategy to use, grade completely on the execution, essentially following the implementation of the strategy guidelines in Version One.

Examples of Scored Student Work

Pages 33–37 show reproductions of actual student work on the Golf Match problem found below and in Chapter 5, Problem Set B, of the student text.

GOLF MATCH

Clark, Chris, Doug, and Diana are standing on the first tee of their favorite golf course, about to begin a best-ball-of-partners match. (A best-ball match pits two golfers against two other golfers.) They are standing in a square, with two partners standing shoulder to shoulder next to each other on the cart path, directly facing the other two partners standing shoulder to shoulder next to each other on the grass. This standing arrangement is typical of the beginning of a golf match. They shake hands, then throw a tee in the air and let it hit the ground. Whoever it points to will tee off first. Clark is standing diagonally opposite Diana. Chris is facing the person whose name begins with the same letter as that of the name of the person who will tee off first. Partners tee off one after the other. Who will tee off second?

The solutions that follow were scored according to the Version One system. You might want to score them first yourself before you read this explanation. You'll find that the scoring is inevitably subjective at times, but if you're consistent, students should understand why they get the scores they do.

Jason and Brandy clearly understood the problem, chose an appropriate strategy (drawing a diagram), carried it out correctly, and stated their answers succinctly and completely. Their explanations are also clear and complete, including explanation of both possible answers. Brandy received a 9 instead of a 10 because she did not name her strategy. So she received only 1 point for the "choosing a solution strategy" section.

Brandy showed a little more of her reasoning by drawing the arrows showing Diana and Clark at diagonals and arrows showing people standing opposite one another. She also mentioned this information in her explanation. Jason explained the partnership relationships. Note that even if you find one of these "perfect" papers better than the other, it should still be possible to score a 10 with a less-than-perfect paper. So you may decide that Brandy deserves a 10 even though she didn't specifically name her strategy.

Don gets only 1 point for understanding. He does understand that Diana and Clark are diagonally opposite each other, and he also understands that Chris is facing the person whose name begins with the same letter as the person who will tee off first. However, he ignores the information in the problem that states partners tee off one after the other. In his explanation he mentions a person who will tee off first and then moves across to the other team for the second person to tee off. Because of this misunderstanding, he gets the wrong answer and thus gets 0 points for the answer. His strategy is appropriate, he carries it out well, and his explanation is good, so he gets full credit in those areas.

Carina draws a diagram showing Diana and Clark diagonal from each other. However, she makes an assumption that Chris cannot be facing Clark because their names start with the same letter. The problem doesn't say this, so she loses a point for understanding. She gets 2 points for choosing an appropriate strategy, but her diagram isn't very good and it accounts for only one possibility, so she gets only 1 point for carrying out the strategy. She does get the right answer, but it's not completely supported—she didn't allow for both possibilities—so she gets only 1 point for her partial answer. She gets full credit for her clear explanation.

Halee's paper is a good example of why you want your students to explain their reasoning in a paragraph or two. It is also a good example of why you need to stress to students that you don't mean scratch work when you say they must show their work. Halee understands the problem and does show diagrams allowing for both possibilities (if you can find them). She correctly reasons through both possibilities to indicate the partnerships and teeing-off order. She also states her answer explicitly (though, despite the stars, it is hard to find at the top of the paper). Thus, she gets points for everything except the missing explanation.

Thanks to Ken Johnson's students Jason Reynolds, Brandy Whittle, Don Elias, Carina Euyen, and Halee Epling for contributing the work copied here.

47

#1
Draw A Diagram

Clark
Chris
Doug
Diana

| Clark > | < Doug |
| Chris > | < Diana |

In this arrangement Chris is facing Diana. This means that Doug tees off first and his partner which is Diana tees off second.

| Clark > | < Chris |
| Doug > | < Diana |

In this arrangement Chris is facing Clark. This means that Chris will tee off first, and his partner Diana will tee of second.

I drew a diagram for each of the possible arrangements and in both of them Diana tees off second.

zzzzz   10

41

Problem Set B
Chapter #5

Brandy W

1. Golf Match

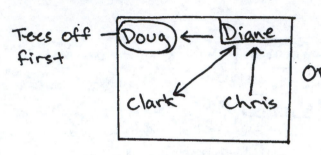 Or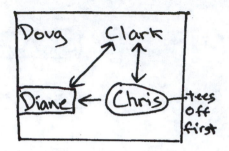

Tees off first →

tees off first

Diane is always second to tee off. No matter who tees off first. Diana is second.

In this instance, Clark is diagonally across from Diane and Chris is across from the person who has the same beginning letter as the person who tees first. Since Diane has the same beginning letter in her name as Doug, then Doug tees off first and Diane second

2¹2²2²
9

In this instance, Diane is diagonally across from Clark. Chris is directly across from the person who has the same beginning letter in his name as the person who tees off 1st. Since Clark is across from Chris, Chris has the same letter "C", therefore Chris tees off 1st and Diane tees off second.

Don E
Math 10
Problem Set B

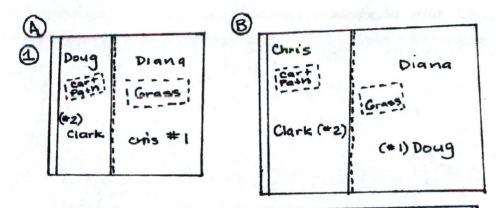

Answer: Clark will always tee off second.

In solving this problem, I decided to draw a diagram (well, actually two diagrams). The _assumption_ I have is is that when two teams have a toss, the players always face their opponents. There are two possible diagrams because we only know half of each team and the "other" half can be with either team. This is why the diagrams simply differ by swapping Chris and Doug. In diagram A, chris is facing Clark who has the same first letter as the person who tee's off first, which is Chris. Then Clark would tee off second. In diagram B, Chris is facing Diana who has the same first letter as Doug who tees off first. Then moving across to the other team, Clark would once again tee-off second. Not to make this too confusing, but Clark always tees off first for his own team.

But in the diagram Diana is Chris' partner!

The problem states partners tee off after one another

12202   7

Carina E

I Golf Match

Clark } Doug
Chris } Diana

Doug will tee off first.
★ Diana will tee off second.

Explanations: I drew a diagram for this problem showing Clark and Diana diagonal to each other. Which left me to figure out who Chris and Diana were teamed up with. From the clue that states that Chris is facing the person whose name starts with the same letter as the person who will start the game, I figured that Chris could not be facing Clark because that would mean that Chris would start first. So that means that Chris is standing opposite Diana. Chris is Clark's partner and Diana and Doug are partners. Therefore since Chris is facing Diana, Doug will start first. And since partners follow each other Diana will go second.

12112

7

(10/5 DUE)

WHO TEES OFF SECOND?

★ DIANE ↺
2nd ★

★Draw a diagram★

Prob Set B pg 136-137   #1-5

① Golf Match:
Names:
CLARK
CHRIS
DOUG
DIANA

2 golfers against two other Golfers.

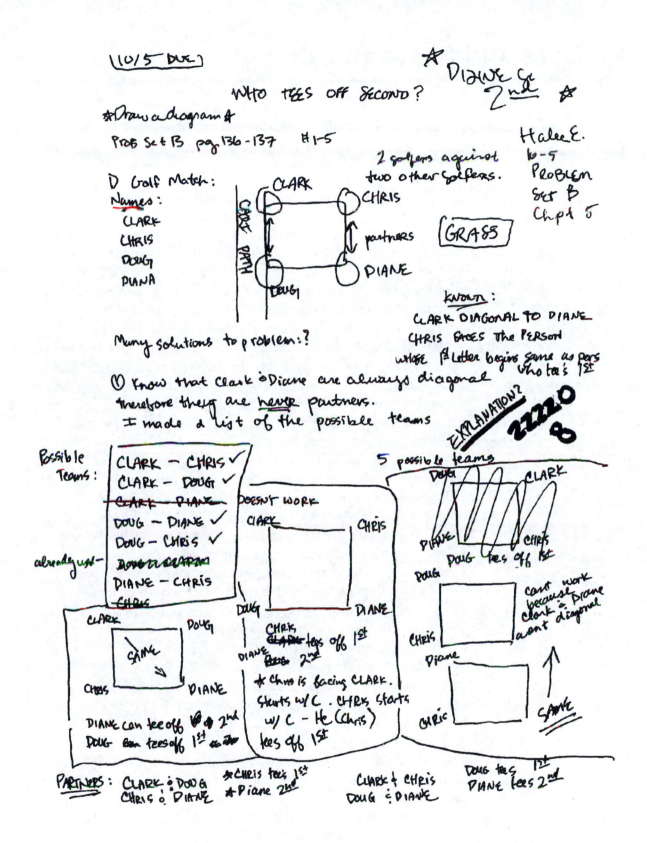

GRASS

Halee E.
10-5
PROBLEM
Set B
Chpt 5

KNOWN:
CLARK DIAGONAL TO DIANE
CHRIS FACES THE PERSON
whose 1st Letter begins same as pers
who tee's 1st

Many solutions to problem:?

① Know that Clark & Diane are always diagonal therefore they are never partners.
I made a list of the possible teams

EXPLANATION? ZZZZ₀ᵒ

Possible Teams:
CLARK — CHRIS ✓
CLARK — DOUG ✓
CLARK — DIANE DOESN'T WORK
DOUG — DIANE ✓
DOUG — CHRIS ✓
already used — DOUG-CLARK
DIANE — CHRIS
CHRIS

CLARK   DOUG
        SAME
CHRIS   DIANE

DIANE can tee off ① & 2nd
DOUG can tees off 1st

CLARK   CHRIS

DOUG   DIANE

CHRIK, CLARK tees off 1st
DIANE tees 2nd
★ Chris is facing CLARK.
Starts w/ C. CHRIS starts w/ C - He (Chris) tees off 1st

5 possible teams

DOUG   CLARK

DIANE   CHRIS

DOUG tees off 1st

DOUG

CHRIS

DIANE

CHRIS   SAME

cant work because Clark & Diane aont diagonal

DOUG tees 1st
DIANE tees 2nd

PARTNERS: CLARK & DOUG   ★Chris tee's 1st
CHRIS & DIANE   ★Piane 2nd

CLARK & CHRIS
DOUG & DIANE

Quizzes and Midterm Exam

Quizzes

We believe that group work results in more effective learning. It is also a realistic representation of what the problem-solving process in most careers looks like: people talking to one another. Individual accountability is important too. Individuals in each group must be accountable to others in any endeavor. Further, students must take responsibility for their education. The most meaningful education takes place when a student decides that education is important and makes the right moves to take charge.

We think that students in the problem-solving course are best served by doing all of the above. Group work and group process are built into most of the course: presentations to the class, working in groups, and so on. The course also emphasizes making choices: deciding when to present a Problem of the Day to the class, selecting among various problem-solving strategies, writing solutions, choosing who to work with outside of class. These two aspects of the course are well represented by the materials given so far in this *Instructor Resources*.

However, one aspect has not yet been discussed: individual accountability. In a group situation, it is possible for a student to allow the other group members to carry him or her along. A student may be able to complete the problem sets by making use of the ideas of other students. While failure to explain solutions thoroughly will hurt a student's grade, we still felt the need for more individual accountability. To ensure this, we recommend giving occasional quizzes on problem-solving strategies.

The quizzes do not need to be difficult. Prepare simple one- or two-problem quizzes at the level of the Problem Sets A, and give the quizzes after students have read the chapter and practiced some problems. If you occasionally choose problems from the text for the quiz, you also reward the students who have taken the time to read the text carefully.

Consider giving a quiz every two or three weeks, starting in week 3, with each quiz covering a sampling of the strategies studied in the preceding weeks. For each quiz, choose no more than two problems (usually one for each strategy) from Problem Set A, Version 2, in this *Instructor Resources*. (Problems that make good quiz problems are indicated in the notes for each chapter.) Version Two of the grading rubrics shown earlier provides a method for scoring the quizzes.

If you require explanations, then they must be very brief; otherwise, the quiz will take away from your class problem-solving time. It is not necessary to give a quiz on every chapter.

This approach builds in individual accountability from the start, especially since the first Problem Set B is not due until the fifth week. Giving short quizzes early in the course establishes the importance of reading the chapter and knowing the material as a standard expectation.

We allow students to use calculators on all quizzes and on the midterm.

Midterm Exam

You may also consider giving a midterm exam. Rather than giving a quiz in the middle of the semester, give a midterm. We recommend giving the midterm at the beginning of week 11, after Chapter 9 has been completed. Because of time constraints, it is a good idea to use problems that ask for incomplete solutions. For example, a guess-and-check problem asks for only three guesses and then stops. Or a subproblems problem might ask for a list of all the subproblems and then stop. Two sample midterms are provided right before the section on final exams.

A Possible Grading Scheme

Course grading can be based on a combination of problem sets, Problems of the Day, quizzes, a midterm, oral presentations, journals, class participation, a final, and whatever else you see as necessary. Here is how one instructor broke down the grading:

- Individual Accountability Component

    Quizzes: 15%

    Midterm: 10%

- Group and Individual Component

    Problem Sets: 45%

    Final: 5%

- Participation Component

    Problems of the Day: 10%

    Oral Presentations: 5%

    Journals: 5%

    Class Participation: 5%

This instructor allowed students to replace their one, two, or three lowest Problem Set B scores with higher scores on Problem Sets B from Chapters 15, 16, and 17.

The POD grade is for completing the Problem of the Day. The oral-presentation grade is for presenting the POD.

The class-participation grade includes points given for questions from the reading and for starting the Problem Set B on the Problem Set B workday. It could also include any problems that students wrote in the "Write Your Own" section of each chapter.

The First Three Weeks and Beyond

Course Schedule

One of this book's authors has taught this course at Sierra College in three different formats. Students receive four units for the course, which fills a 17-week semester, including 1 week set aside for final exams. The course meets 4 hours per week, broken up in one of these ways:

a. 4 days per week at 50 minutes per class period

b. 2 days per week at 1 hour 50 minutes per class period

c. 3 days per week at 1 hour 15 minutes per class period

In addition, he taught the course in a 5-week summer school, meeting 4 days per week at 3 hours 20 minutes per class period.

To finish most of the book, consider adopting a schedule similar to the following one.

Week 1

On part of the first day, do some group-building activities—possibly the jigsaw puzzle or interview activity and the Color Square Game. Assign two Problems of the Day (PODs) from the Introduction such as Soccer Game, Elevator, and Ducks and Cows.

On the second and third days, have students make presentations and maybe do another group-building activity. Begin Chapter 1 by introducing diagrams and having students work in groups on problems from Problem Set A. For a POD, give another problem from the Introduction and one from Chapter 1, such as Floor Tiles.

Week 2

Finish the material from Chapter 1 by having students work in groups on other problems from Problem Set A such as Stone Necklace and Dangerous Maneuvers. Assign the Haywire problem as a POD. Begin Chapter 2. Model a systematic list, probably the Loose Change problem from the text. Model a couple of different systems. Have students work in groups from Problem Set A. Good PODs are Storage Sheds, Basketball, and Tanya's Terrific T-Shirts.

If you didn't quite finish Chapter 2, then spend a little more time on it. Regardless, start Chapter 3. At this point, begin a schedule of one new chapter per week. So week 3 is Chapter 3, and so on. A good introduction to the strategy of eliminate possibilities is conundrums, described in this book in the section about Chapter 3.

In all chapters, class time generally should be spent on student presentations (for about 20 minutes) and most of the rest of class should be spent on students working in groups using the new strategy to solve problems. The instructor should not spend much time talking to the class.

At the end of the week, give the NELSON + CARSON = REWARD problem for the groups to solve. Then assign students to do a sample write-up of the problem over the weekend. This introduces problem set write-ups.

You may also want to give a quiz this week—featuring one diagram problem and one systematic list problem. Good PODs for Chapter 3 are Eggs in a Basket, Wow, Wow, So Cook!, and To Tell the Truth.

Work through Chapter 4. Model one problem, illustrating the strategy of matrix logic. Then, again, assign students to do problems in groups, practicing the new strategy. You should also cut back on assigning PODs this week, as Problem Sets B will start.

Assign Problem Set B from Chapter 3 this week, to be due the following week. Sometime late in the week, give the students a significant amount of class time to work on Problem Set B in their groups. If your class meets only 3 hours per week, you can probably afford to give them only 30 minutes to work together. If your class meets 4 or 5 hours per week, then you should be able to give them 45 to 50 minutes to work together.

The schedule for all the other weeks should be similar to the schedule for week 4:

- Students learn new material from Chapter X. Have them do group work on the new strategy, as well as one or two PODs from the new strategy.

- Students turn in Problem Set B from Chapter $X - 2$.

- Assign students Problem Set B from Chapter $X - 1$, and provide some class time to work on it in groups.

- Possibly quiz on one or two previous strategies. Don't give quizzes every week. Design quizzes to take no more than 30 minutes.

For problem-set workdays, consider requiring the students to have started each of the problems in the problem set. Give one point for each problem started, and count that toward their class-participation grade. Requiring this work ahead of time will make the in-class groups much more effective—and also will force students to be more accountable.

If your class meets two days per week, either MW or TT, we suggest making the problem set due on the first day and allowing time to work on the next problem set on the second day. If your class meets MWF, you could make the problem set due on Wednesday and allow time to work on the next problem set on Friday.

Chapters 1–14 form the core of the book. If that is all you are able to cover, your students will come away with a lot of new strategies. However, if you have time remaining, consider selecting problems from Chapters 15, 16, and 17. You could assign Problem Set B from each of these chapters as extra-credit work, due on the day of the final exam. The scores on these problem sets can be used to replace low scores on earlier problem sets. Students who have successfully completed the earlier problem sets do not have to do these.

Adjustments for a Three-Unit Course

All of the preceding material describes a 4-hour-per-week course. If your course meets only 3 hours per week, you will obviously have to make some adjustments. You will probably not have enough time to devote a full class hour to students working on Problem Sets B. You should still give them some time, but maybe only 25 to 30 minutes. You may also want to give fewer presentation opportunities. Experiment to find what works best for you. You may not be able to get through the whole book, but you should try to get through Chapter 14.

Teaching Resources for Chapters 1 through 17

Draw a Diagram

Drawing a diagram is a central problem-solving skill. It is applied in many disciplines, both academic and vocational. Drawing a diagram is a form of organizing information in a spatial way. This is a strategy to use to organize your thoughts or to explain something to someone else. As the text mentions, a picture is worth a thousand words, and even that may represent a serious undercount. Much information, such as this page of text, is presented in a linear, one-dimensional fashion. The reader is expected to start at the top left and read to the right on each line. If not for the edge of the page, the text could continue infinitely in a line. But a diagram is different. The viewer can start anywhere and proceed in any direction within the diagram. The change from a linear presentation to a two-dimensional one frees up the information being presented. The extra dimension can convey relationships among pieces of information spatially. Organizing information spatially allows the visual-processing portion of your brain to become involved in problem solving.

Our sense of vision is extremely well developed, both biologically and by training. We're called on more and more to interpret visual images: Newspapers feature graphic communication; computers operate with interfaces that are primarily graphic. The question today is not whether our electronic gadgets should have graphical interfaces but, rather, how those interfaces should be organized and presented.

There are times when diagrams serve as a main communication medium and other times when they play a supporting role. For example, in a newspaper article, a diagram may show a parade route. The diagram could actually be the main communication if the article is simply about the parade route. On the other hand, if the article talks about more than the parade route, the diagram plays a supporting role.

Your students may resist drawing diagrams. In many cases, students' prior problem-solving experience is algorithm- or equation-based. To motivate students to learn to draw diagrams, point out that many people make their living in graphics. For example, graphic artists create images that convey information or feelings.

TEACHING RESOURCES

©2004 Key College Publishing, *Instructor Resources: Crossing the River with Dogs,* Johnson/Herr/Kysh

A student who resists learning a strategy may miss a significant number of otherwise easy problems. Fortunately, the resulting cognitive dissonance should cause the student to reexamine his or her approach and adjust. Written work can document to a student that his or her approach or attitude toward a strategy isn't working.

In these early chapters, emphasize to students that they need to write complete solutions with explanations. One of the emphases of this book is to develop mathematical communication skills. Students who write down their work and explanations will find it easier to recognize incorrect solutions and determine where they went wrong.

Notes on Text Problems

VIRTUAL BASKETBALL LEAGUE

This is a rich problem that can be solved by a number of different strategies. At some point later on, ask students to find as many ways as they can for solving this problem. It will be discussed again in Chapter 2: Make a Systematic List.

MODEL TRAIN

Students will tend to try to solve this problem in their heads because it appears so simple. They should be writing down their work on this for two reasons: to practice explaining their work and to be accountable for their processes and solutions.

FARMER BEN

This is another rich problem that can be solved in several different ways. A similar problem, the Farmer Jones problem, appears in Chapter 6: Guess and Check and in Chapter 13: Convert to Algebra.

Notes on Problem Set A

WORM JOURNEY

Some groups may come to the conclusion that 1 foot gained each day means that it takes 12 days. Be sure they draw the diagram.

UPS AND DOWNS OF SHOPPING

There is an unstated assumption in the problem that the credit department is on the middle floor. If students are getting the incorrect answer of 11 floors for this problem, ask the question "What does it mean to be in the middle?" or "Is the TV department on the top floor?"

Notes on Problem Set A *(continued)*

To emphasize the idea of being in the middle, point to three students that are in a line and ask who is in the middle. Do the same with four students, and then five students.

DANGEROUS MANEUVERS

Encourage students to make a two-dimensional map. It is much harder to put all the camps in one straight line in the diagram. Note that the shortest way from Lupine to Bovine is 10 miles going through Canine first, rather than traveling directly from Lupine to Bovine. While this may appear to contradict the triangle inequality, ask students how this could occur. Most students will come up with ideas such as that the road from Lupine to Bovine is really winding or that it goes over a hill.

HAYWIRE

This is a good Problem of the Day. For the student presentations, it gets boring if everyone describes what they did and goes through the answers to all of the questions in the book. Instead, it is a good idea to have each presenter put just a diagram on the board. Then each presenter can simply answer a question, "How would A call B?" You can ask any questions you want—don't be limited by the eight questions in the text. It's a good group-building exercise to have each presenter pose a question to the next presenter.

Good Quiz Problems

For a one-problem quiz on this strategy, consider one of these problems from Problem Set A, Version 2: 9, Malfunctioning Rototiller; 10, Rundown; 17, Vacuum Cleaner; 18, Campaigning.

Notes About Questions from the Reading

When you read a math book, you must stop and think about virtually everything you read. Very few people are used to this technique. The Questions from the Reading sections are designed to help the students develop the technique. Questions from the Reading will appear only for the first three chapters of the textbook. The following pages have been formatted to be copied but with the intent that the students do their work on another sheet of paper—the pages are not formatted to be used as fill-in-the-blank worksheets.

You may wish to have the students work on the questions individually at home, in pairs, or as a group. You may also decide to do the questions orally in a whole-class discussion.

Questions from the Reading

On a separate piece of paper, answer each question with a full sentence or sentences. Use the information from your textbook to answer each question.

1. List some synonyms for the word *diagram*.

2. List some occupations that use diagrams.

VIRTUAL BASKETBALL LEAGUE

3. What do the points and lines represent in Rita's diagram?

4. What do the numbers 21 and 3 represent in Rita's solution?

MODEL TRAIN

5. List the names of the people who solved this problem and give a brief description of each one's solution method.

6. Explain the confusion about whether the train had gone one-half or one-third of the way around the track.

7. Draw a diagram of the solution that gave "25 seconds" as the answer. Describe what went wrong in this solution.

THE POOL DECK

8. Rajesh did this problem in three parts. Three of the diagrams below illustrate these parts, and the other diagram illustrates part of someone else's approach. Identify whose approach corresponds to each diagram.

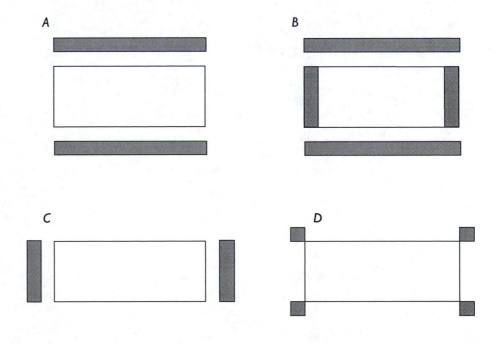

A

B

C

D

Questions from the Reading *(continued)*

9. Draw a diagram with the corners attached to the width, and calculate the area of the deck using your diagram.

10. Explain why Herb's approach to this problem works. Draw a diagram that Herb may have used.

FARMER BEN

11. What sort of biological engineering did Trent do?

CONCLUSION: DRAW A DIAGRAM

12. Fill in three bullet points for this phrase:
"Diagrams are often necessary to show . . ."

- _____
- _____
- _____

Answers to Questions from the Reading

1. Many answers are possible: drawing, picture, schematic, map, blueprint, graph, outline, plan, illustration, representation, sketch, chart.
2. Many answers are possible, including law enforcement, architecture, sports, landscape, marketing, transportation, graphic design, construction.

VIRTUAL BASKETBALL LEAGUE

3. The points represent teams and the lines represent matchups (a game).
4. Twenty-one is the number of different pairs of teams playing a game, and three represents the number of games each pair plays against each other.

MODEL TRAIN

5. Rena divided the track into thirds and got 10 seconds for each third, for a total of 30 seconds.
   Phong divided the track into sixths and got 5 seconds for each sixth, for a total of 30 seconds.
   Pete also divided the track into sixths but failed to have the train go from the sixth to the first pole and thus got only 25 seconds.

6. Many answers are possible: Confusion resulted because going from the first to the third pole can sound like three of six poles have been passed. However, it covers only two out of six intervals between the poles, hence it is one-third of the track.
7. Answers will vary. Students should draw a picture showing six poles and the train going from pole 1 to pole 6, and not returning to pole 1.

THE POOL DECK

8. Rajesh used diagrams A, C, and D. May used diagram B.
9.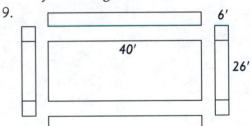

$$26 \text{ ft} \times 6 \text{ ft} = 156 \text{ ft}^2$$
$$156 \text{ ft}^2 \times 2 = 312 \text{ ft}^2$$
$$40 \text{ ft} \times 6 \text{ ft} = 240 \text{ ft}^2$$
$$240 \text{ ft}^2 \times 2 = 480 \text{ ft}^2$$
$$312 \text{ ft}^2 + 480 \text{ ft}^2 = 792 \text{ ft}^2$$

10. Answers will vary. Herb's diagram would look something like this:

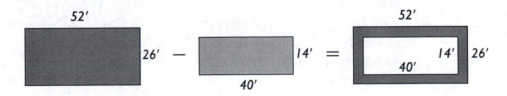

FARMER BEN

11. He converted ducks into cows.

CONCLUSION: DRAW A DIAGRAM

12. Position, directions, complicated multidimensional relationships.

©2004 Key College Publishing, *Instructor Resources: Crossing the River with Dogs*, Johnson/Herr/Kysh

Text Problems

VIRTUAL BASKETBALL LEAGUE

Andrew and his friends have formed a fantasy basketball league in which each team will play three games against each of the other teams. There are seven teams: the (Texas A&M) Aggies, the (Purdue) Boilermakers, the (Alabama) Crimson Tide, the (Oregon) Ducks, the (Boston College) Eagles, the (Air Force) Falcons, and the (Florida) Gators. How many games will be played in all? Do this problem before reading on.

MODEL TRAIN

Esther's model train is set up on a circular track. Six telephone poles are spaced evenly around the track. The engine of Esther's train takes 10 seconds to go from the first pole to the third pole. How long would it take the engine to go all the way around the track? Solve the problem before reading on.

THE POOL DECK

Curly used a shovel to dig his own swimming pool. He figured he needed a pool because digging it was hard work and he could use it to cool off after working on it all day. He also planned to build a rectangular concrete deck around the pool that would be 6 feet wide at all points. The pool is rectangular and measures 14 feet by 40 feet. What is the **area** of the deck? As usual, solve this problem before continuing.

FARMER BEN

Farmer Ben has only ducks and cows. He can't remember how many of each he has, but he doesn't need to remember because he knows he has 22 animals and that 22 is also his age. He also knows that the animals have a total of 56 legs, because 56 is also his father's age. Assuming that each animal has all legs intact and no extra limbs, how many of each animal does Farmer Ben have? Do this problem, and then read on.

Problem Set A, Version 2

1. THE GIFT EXCHANGE

Eight relatives always give gifts to one another. They've been doing it annually now for five years. How many gifts have been given so far?

2. AIR WARBUCKS

There is a rumor that Daddy Warbucks is starting a new airline. He plans to serve six cities: San Francisco, Los Angeles, Dallas, Chicago, New York, and Miami. There will be direct flights between each pair of cities. How many different routes will this create?

3. COOKIE EXCHANGE

Nine people at a cookie exchange each brought a dozen cookies for each other person. How many cookies were brought to this exchange?

4. THE FINANCE COMMITTEE

Nine members of a finance committee could not get together for an emergency meeting. However, during the next three days, each member talked to each other member on the telephone. What is the minimum number of phone calls needed to accomplish this?

5. THE MERRY-GO-ROUND

On a merry-go-round, 12 horses are evenly spaced on the outside perimeter. It takes four seconds for the fourth horse to reach Abe, starting when the first horse is opposite him. How long will it take the merry-go-round to go around once?

6. CREEPING UP IN AGE

Allen, Lydia, Cindy, and Marcus are all friends. Marcus says he is older than Cindy, and he should know because he is her brother. Allen knows he's older than all of them, and Cindy is pretty sure that Lydia is younger than she. List the friends in order from oldest to youngest.

7. CARISSA'S TABLE

Carissa is building a new dining-room table. The center of it is a rectangular section made from polished rocks. The section measures 36 inches by 64 inches. She will place a wooden, eight-inch rectangular lip all the way around the perimeter. What is the area of the wooden part?

©2004 Key College Publishing, *Instructor Resources: Crossing the River with Dogs*, Johnson/Herr/Kysh

Problem Set A, Version 2 *(continued)*

8. R-DOUBLE-7 RANCH

The R-Double-7 Ranch has a new owner. The 30 animals, all ostriches and horses, are dismayed because they have heard that he is both foolish and inexperienced. This turned out to be true—not being quite sure what he was looking for, he checked on the health of his animals by inspecting all of their feet. There were 96 feet in all. How many ostriches are there on this ranch?

9. MALFUNCTIONING ROTOTILLER

The gearshift on Maxine's rototiller was malfunctioning. She found that it would travel forward 4 feet in one minute, but then shift into reverse and go back 3 feet before the shift would allow her to put it back in forward. The backwards part also took about a minute to complete. How long would it take her to till a 13-foot-long section of her garden with this rototiller?

10. RUNDOWN

We told Gavin not to run, but he did anyway. He took about 18 steps from second base, got caught in a rundown, took 7 steps back, then took 3 steps forward, 5 steps back, 11 steps forward, 4 steps back, and was tagged out halfway between second and third base. How many steps is it from second base to third base?

11. THE DOG WHO LIKES TO RUN

Annie the Dog loves to run. When we were walking home one day, she ran ahead of us, went all the way home, and then came back to meet us. She then ran back home again and came back to meet us. She did this three more times. Each time she came back, she met us at a point halfway from where she had left us. If she started running while we were about half a mile from home (let's say about 2400 feet), how far did she run in total by the time she met us the fifth time?

12. TRAVELING ART SHOW

Ardith was in charge of making motel arrangements for the crew of a traveling art show. She had misplaced her list of cities, though, and needed to figure out how many more cities there were. She remembered that the 8th city was Phoenix and the 17th city was Pensacola, and that the cities from Phoenix to Pensacola, inclusive, made up one-third of the tour. How many cities are there on this tour?

Problem Set A, Version 2 *(continued)*

13. **GEORGETOWN RACE**

Five cars remained in contention by the end of the Georgetown One-Hundred-Mile Classic-Car Race. The Duesenberg finished 12 seconds ahead of the Edsel. The Studebaker finished 7 seconds behind the Model T Ford. The time from when the first finisher crossed the line to the last finisher was 22 seconds. The Pierce-Arrow finished 14 seconds after the Model T Ford. The Edsel came in third place. List the cars in the order in which they completed the race, and the times between them from the first to fifth places.

14. **CLASSROOM CEILING**

A classroom measures 24 feet by 36 feet. Each of the panels for the hung ceiling measures 1½ feet by 4 feet. How many panels will be required for the entire hung ceiling?

15. **ARKANSAS CITIES**

According to a map of Arkansas, a number of good roads connect some of the towns south of Little Rock. The road between Malvern and Benton is 23 miles long, and between Benton and Pine Bluff is 62 miles long. The road between Malvern and Sheridan is 28 miles long, and then from Sheridan to Pine Bluff is 22 miles long. Arkadelphia is connected to two towns: It is 12 miles from Gurden and 21 miles from Sheridan. Fordyce has three roads to these towns: 45 miles to Pine Bluff, 39 miles to Sheridan, and 27 miles to Camden. The road from Gurden to Camden is 38 miles.

Unfortunately, the road from Fordyce to Sheridan is closed due to dangerous road conditions caused by the recent bad weather. Find the shortest route between each pair of towns:

Malvern and Pine Bluff

Camden and Pine Bluff

Benton and Arkadelphia

Fordyce and Malvern

Benton and Camden

©2004 Key College Publishing, *Instructor Resources: Crossing the River with Dogs*, Johnson/Herr/Kysh

Problem Set A, Version 2 *(continued)*

16. **WILDERNESS CAMP**

At a wilderness survival training camp, each person was instructed to go out individually, set up camp, and survive for one week. People got points for pinpointing anybody else's camp on a map. You lost points if other people were able to find your camp. About three days into the exercise, Adrienne had come upon both Freda's and DuJannie's camps. Bart had found Eugenia's and Glenda's camps, and he and DuJannie had found each other's camp. DuJannie had also happened upon Hank's camp, while Hank and Eugenia had found each other's. Hank had also discovered Adrienne's. Carl had detected Glenda's camp, and his camp had been detected by Freda.

Unfortunately, the base leaders needed to get in touch with Eugenia because someone had crashed into her parked car. They found Carl. How could they get a message to Eugenia?

Suppose Hank decides to send a message to the leaders the next day. How could he get it out by way of Carl (and, of course, without any more camps being discovered)?

How could you get messages from

a. Freda to DuJannie?

b. Adrienne to Glenda?

c. Glenda to Hank?

d. DuJannie to Freda?

e. Bart to Carl?

f. Carl to Bart?

17. **VACUUM CLEANER**

Janae was vacuuming the narrow hallway in her house. She went 5 feet forwards in the first 4 seconds, then went 3 feet backwards in the next 4 seconds. She continued to do this, forwards 5 feet in 4 seconds and backwards 3 feet in 4 seconds. If she continues in this way, how many seconds will it take her to reach the end of her hallway, which is 15 feet long?

TEACHING RESOURCES

©2004 Key College Publishing, *Instructor Resources: Crossing the River with Dogs*, Johnson/Herr/Kysh

18. **CAMPAIGNING**

Dean was campaigning for president. Supporters were lined up along one side of a walkway hoping to shake his hand. He started at one end of the walkway, walked forwards 6 yards, and stopped to shake hands with James. Then he walked backwards 2 yards and shook hands with Linda. Then he walked forwards 7 yards and shook hands with Nicole. Then he walked backwards 3 yards and shook hands with Rebecca. Then he walked forwards 8 yards and shook hands with Tadashi. Then he walked backwards 4 yards and shook hands with Marisa. Finally, he walked forwards 6 yards to the end of the walkway.

a. How long is the walkway?

b. How far apart are Tadashi and James?

2 Make a Systematic List

Systematic-list problems will give you some opportunity to assess students' learning from previous courses. Although systematic lists are relatively simple, watch for students who can do them well, because they have probably had experience with them. Usually you will find that people can make the list, but how *systematic* is the list? Students may resist this strategy at first and need guidance developing functional systems for their lists.

Systematic lists are a way of organizing information. The opposite approach, often called "random flailing," does not help people function well. Organization makes things easier on the brain because the brain organizes information while storing it. Giving information a structure also allows the brain to absorb and interpret the information more easily.

A systematic list can be made relatively quickly and efficiently. It can also be verified quickly by both the writer and the reader. Patterns will emerge from the list. Patterns promote further insight and serve as verifiers when checking for accuracy. When a list is done systematically, it becomes more of a learning tool.

For students using this book, communication must be a high priority. Part of the problem-solving process is to communicate the solution to a problem. To communicate a solution, you must understand it and record it in a form that is understandable.

Emphasize that students should work slowly and carefully. Careful work is especially important as students become evaluators of their own and their peers' work. Furthermore, concise solutions will be far easier to communicate than rambling solutions.

Students also need to realize that if a method does not work well the first time they try it, they may need to try a different method. Rather than continuing blindly down a dead end, students need to back out and find a better avenue. Starting over on a problem is a necessary skill; when someone starts over, she is starting again at an increased knowledge level—she's already learned what doesn't work in solving a particular problem.

Systematic lists show up in a number of places—so many that we may not think of them as systematic lists at first. The phone book is a prime example: It is an alphabetical listing of people and organizations. Its systematic nature makes it easy to use.

As your students progress through the book, you will find more and more problems that can be approached in a variety of ways. To encourage divergent thinking, you must value alternative approaches, even if some are not necessarily clear or efficient. With some problems, one method may not emerge as being the most efficient; instead, different methods will exhibit different strengths.

Notes on Text Problems

VIRTUAL BASKETBALL LEAGUE

This rich problem reappears to be solved by a different method. The problem can be solved in a number of ways, and a number of different methods can be used to make the systematic lists.

PENNY'S DIMES, PART I

An important assumption in this problem rests on whether or not you can distinguish between the piles (and care that you can). The solution presented in the text assumes a case in which you do not distinguish.

Notes on Problem Set A

The first few problems—through Making Change—are not difficult. The rest of the problems are difficult. In Problem Set A, Version 2, Fencing with Neighbors and Arcade are very difficult.

STORAGE SHEDS

This problem makes an excellent Problem of the Day. A good discussion should ensue about whether or not an 8-by-10 shed is the same as a 10-by-8 shed. Students should give several arguments on both sides. Do not state an opinion on the matter, just listen to the discussion.

FINISHED PRODUCT

This problem is a little more difficult. Suggest that students start their list with 1, 360, then 2, 180, and so on. Even though the first few pairs have a sum of more than 100, including them helps students organize their list. These pairs can be crossed out later.

Notes on Problem Set A *(continued)*

■ TWENTY-FOUR

By this time, many students may be used to using a chart format, like the solutions to Loose Change on pages 26 and 27 of the text. However, a list that starts like the one below is much easier to do.

> $2 + 2 + 2 + 18$
> $2 + 2 + 4 + 16$
> $2 + 2 + 6 + 14$
> $2 + 2 + 8 + 12$
> $2 + 2 + 10 + 10$

The basic rule of a list like this is that each entry should have numbers that do not decrease. Thus, when you get to $2 + 2 + 12 + 8$, that must be a repeat of an earlier entry ($2 + 2 + 8 + 12$). So the list would continue with $2 + 4 + 4 + 14$.

■ TARGET PRACTICE

Again, a chart format like Derrick's Frisbin solution on page 32 of the text does not work as well here. Suggest to your students that they try to do this problem in the way Julian solved Frisbin on page 33.
Start the list:

25	25	25	25
25	25	25	10
25	25	25	5
etc.			

■ TANYA'S TERRIFIC T-SHIRTS

This problem makes an excellent Problem of the Day. It is very difficult. There are essentially two ways to do this problem:

1. Solve the problem for $40, then for $35, then for $30, . . . all the way down to $5.

2. Make a list that contains all the different dollar amounts mixed together.

Good Quiz Problems

For a one-problem quiz on this strategy, consider one of these problems from Problem Set A, Version 2: 11, Making Change for 70 Cents; 12, Making Change for 55 Cents. (Note that any amount of change between 55 and 75 cents works well.)

Questions from the Reading

On a separate piece of paper, answer each question with a complete sentence or sentences. Use the information from your textbook to answer each question. Questions with "explain" or "compare and contrast" need at least a well-formed paragraph and possibly examples.

LOOSE CHANGE

1. Compare and contrast Brooke's and Heather's systems for organizing a list of possibilities.

VIRTUAL BASKETBALL LEAGUE

2. What does AB mean in Monica's list?

3. Why doesn't Monica include BA in her solution?

4. What are some patterns that help verify that Monica has completed her systematic list correctly?

PENNY'S DIMES, PART I

5. Explain what Randy's decision was about.

6. What is another possible way to set up a systematic list for this problem?

7. Tyson solved the problem by using a systematic list that was identical to Randy's. Then Tyson said, "I wanted to know how many ways there would be if you could tell the difference between the piles. So I multiplied 16 times 3 and got 48 ways." Discuss Tyson's reasoning. Do you agree?

FRISBIN

8. Summarize Derrick's method for solving this problem.

9. Describe the pattern in the 5 points column of Derrick's solution.

WHICH PAPERS SHOULD KRISTEN WRITE?

10. Briefly compare and contrast Li's and Travis's methods for solving this problem.

Answers to Questions from the Reading

LOOSE CHANGE

1. Many answers are possible. Brooke and Heather both created systematic lists and used only pennies, nickels, and dimes as per the problem statement. Both listed possibilities by going from larger to smaller numbers. Their systems differ in that Brooke started with the largest valued coins (dimes), first finding all combinations of two dimes, then of one dime, and then of zero dimes. Heather, on the other hand, started with the greatest number of coins (pennies) and found all possibilities for 25 pennies, for 20 pennies, and so on until she had exhausted all possibilities.

VIRTUAL BASKETBALL LEAGUE

2. It stands for Aggies versus Boilermakers.

3. It is already taken care of with the AB matchup.

4. Answers will vary. Some possible answers are (a) the visual pattern of the columns, (b) the pattern of descending number of matchups from left to right, (c) the patterns of the letters within the solution, both vertically and horizontally.

PENNY'S DIMES, PART I

5. Randy had to decide if 1, 11, 13 was the same as 1, 13, 11.

6. You could start with 23 dimes in the first pile.

7. Tyson's answer is incorrect. There would be six ways to arrange each set of three. Example: If the numbers were 1, 3, 21, they could be arranged as 1, 3, 21; 1, 21, 3; 3, 1, 21; 3, 21, 1; 21, 1, 3; 21, 3, 1. That would give $16 \times 6 = 96$ ways. However, if there were a repeat in a pile, like 7, 7, 9, there would be only three ways to rearrange the numbers: 7, 7, 9; 7, 9, 7; 9, 7, 7. This happens with double 1's, 3's, 5's, 7's, 9's, and 11's. Each double contributes 3 fewer ways, and $6 \times 3 = 18$, which makes the total $96 - 18 = 78$. This is the same answer you would get if you wrote out all the numbers.

FRISBIN

8. Derrick started with the largest number of 10-point throws, then showed all possibilities that had two 10-point throws. He then did the same for all possibilities that had one 10-point throw, and finished off with all possibilities for zero 10-point throws.

9. It is a series of sets of descending numbers. Each set has one more number in it than the previous set. For example, the first set is one number: 0. The second set is two numbers: 1, 0. The third set is three numbers: 2, 1, 0. This pattern continues.

WHICH PAPERS SHOULD KRISTEN WRITE?

10. They both used systematic lists. They both exhausted all possibilities of *Richard III* before moving on. They both ended up listing all the books in the same order. They differ in that Li organized her solution as a list, and Travis organized his solution as a check-off chart.

©2004 Key College Publishing, *Instructor Resources: Crossing the River with Dogs,* Johnson/Herr/Kysh

Text Problems

LOOSE CHANGE

Leslie has 25¢ in her pocket but does not have a quarter. If you can tell her all possible combinations of coins she could have that add up to 25¢, she will give you the 25¢. Solve this problem before continuing.

VIRTUAL BASKETBALL LEAGUE

Andrew and his friends have formed a fantasy basketball league in which each team will play three games against each of the other teams. There are seven teams: the (Texas A&M) Aggies, the (Purdue) Boilermakers, the (Alabama) Crimson Tide, the (Oregon) Ducks, the (Boston College) Eagles, the (Air Force) Falcons, and the (Florida) Gators. How many games will be played in all? Do this problem before reading on.

PENNY'S DIMES, PART I

Nick's daughter Penny has 25 dimes. She likes to arrange them into three piles, putting an **odd number** of dimes into each pile. In how many ways could she do this? Solve this problem before continuing.

FRISBIN

On a famous episode of *Star Trek,* Captain Kirk and the gang played a card game called Phisbin. This problem is about another game, called Frisbin. The object of Frisbin is to throw three Frisbees at three different-sized bins that are set up on the ground about 20 feet away from the player. If a Frisbee lands in the largest bin, the player scores 1 point. If a Frisbee lands in the medium-sized bin, the player scores 5 points. If a Frisbee lands in the smallest bin, the player scores 10 points. Kirk McCoy is playing the game. If all three of his Frisbees land in bins, how many different total scores can he make? Make a systematic list for his problem before reading on.

Text Problems (continued)

AREA AND PERIMETER

A rectangle has an area measuring 120 square centimeters. Its length and width are whole numbers of centimeters. What are the possible combinations of length and width? Which possibility gives the smallest perimeter? Work this problem before continuing.

©2004 Key College Publishing, *Instructor Resources: Crossing the River with Dogs,* Johnson/Herr/Kysh

WHICH PAPERS SHOULD KRISTEN WRITE?

For her Shakespeare course, Kristen is to read all five of the following plays and choose three of them to write papers about: *Richard III, The Tempest, Macbeth, A Midsummer Night's Dream,* and *Othello.* How many different sets of three books can Kristen write papers about? Do the problem before continuing.

©2004 Key College Publishing, *Instructor Resources: Crossing the River with Dogs,* Johnson/Herr/Kysh

Problem Set A, Version 2

1. RIDE TICKETS

Marcy had $10.00 to spend at the state fair. She spent all of it on ride tickets. The rides cost $0.50, or $1.00, or $1.25 each. What are all the possible ways in which she spent her money?

2. LEARNING THE HARD WAY

Rhoda, Quincy, Sen, and Trey went boating. Sen and Trey got into an argument. They stood up, then Rhoda and Quincy both stood up and told them to stop and sit down. One by one, each person fell off the boat. List all the possible orders in which they could have fallen.

3. RENTING A CAR

Axxel Car Rentals has two basic plans. One is to rent a car for $19 per day and then pay $0.20 per mile. The second is to rent a car for $49 for the first day and then pay $29 for each subsequent day with unlimited mileage. Figaro expects to drive a car about 500 miles total, but he has no idea how many days he'll need it. (He might need it for two days, but he could see needing it for as many as ten days.) Make a systematic list to help him decide which plan to use.

4. WAYLON'S CANDY

The corner convenience store sells candy in 20¢, 30¢, and 50¢ packages. List all the ways in which Waylon can spend exactly $3.00 on candy.

5. FENCING WITH NEIGHBORS

Seenonaybor's Fencing Company makes prefabricated fence units in 6-, 8-, 10-, and 12-foot lengths. What combinations could be used to make exactly 30 feet of a straight fence?

6. ARCADE

Kerry challenges Siri at the arcade. They each will roll three balls and then total their scores. Whoever has the higher score wins. The possible scores for the holes the balls can fall into are 50, 40, 30, or 25 points, and every ball falls into a hole. What are the possible scores for three balls?

7. RUDY'S SHOT TOTALS

In the first quarter of a basketball game, Rudy scored 8 points on a combination of 3-point field goals, 2-point field goals, and 1-point free throws. What are all the possible combinations of shots that could have scored 8 points?

©2004 Key College Publishing, *Instructor Resources: Crossing the River with Dogs*, Johnson/Herr/Kysh

Problem Set A, Version 2 *(continued)*

8. SKIING BIKERS

The bicycling club rented three vans to take people skiing. Each van could hold 7 people. As it turned out, only 12 people could make the trip, but because of the amount of equipment they had to bring, they still needed all the vans. Peter, the leader, said, "I don't care who goes in what van, but obviously we need at least one driver in each van." Without regard to which vans the three groups get into or to who is in which group, in how many different ways can the 12 people be split up?

9. SUBJECT-VERB-OBJECT

A common structure for English sentences is subject-verb-object. An example is "The horse pulled the cart." *Horse* is the subject, the verb is *pulled*, and *cart* is the object. Here is a list of verbs and nouns. The nouns may be used as subjects or as objects.

NOUNS		VERBS
alligator	bone	eat
dog	bird	scare
car	soda	drive
wrench		pour

How many different subject-verb-object sentences can be created with the words in the lists? For example, you can create the sentence "The bone eats the dog" by using *bone* as the subject, *eat* as the verb, and *dog* as the object. As you can see, the sentences you create may be ridiculous. The same noun cannot be used more than once in a given sentence.

Suppose you were going to write all the possible sentences with the words in the lists. Devise a method for dividing the work among the members of your class.

10. MAKING CHANGE FOR 70 CENTS

How many ways are there to make change for 70 cents, using quarters, dimes, and nickels? *Note:* **No pennies!**

11. MAKING CHANGE FOR 55 CENTS

How many ways are there to make change for 55 cents, using quarters, dimes, and nickels? *Note:* **No pennies!**

©2004 Key College Publishing, *Instructor Resources: Crossing the River with Dogs,* Johnson/Herr/Kysh

3 Eliminate Possibilities

Eliminating possibilities is a reversal of sorts. You are not looking for the right answer; rather, you are looking for a lot of wrong answers. As each wrong answer is eliminated, you get closer to the right answer (or answers).

Eliminating possibilities is useful in any sort of troubleshooting. Many times a problem exists but is not apparent. To find out what it is, you eliminate whatever is *not* wrong, and whatever is left over must be the problem. This is one of the reasons why students must write out complete lists of all possibilities. If a student's incomplete list doesn't include the correct answer, he or she may eliminate all the answers on the list and be left with the incorrect impression that the problem has no right answer.

Many mathematics students think problems have only one answer. This is not always true but is an impression students can get from traditional mathematics courses. Lately, more mathematics curricula are becoming more integrated with other subjects. Mathematics is taught and used more as a thinking and analytical tool than as an isolated academic discipline. In this type of setting, "correct" answers become more difficult to ascertain. In this text, many problems have more than one correct solution.

Occasionally, we must make up possibilities in order to have something to eliminate. The Who Is Lying? problem in this chapter requires this process. The text refers to this process as *seeking contradictions*. Seeking contradictions involves making up a possibility and then eliminating it to show that its opposite is true. This sort of process is similar to that used for an indirect proof.

Playing Twenty Questions is a good way to introduce the strategy of eliminating possibilities. One way to play the game is described in the text: Choose a number between 1 and 100, and have students ask yes-or-no questions to determine the number. Another, more interesting, way to play the game is to choose an item in the room. The item can be a specific piece of paper on the wall, a person, a piece of chalk, the clock, anything. The class again asks yes-or-no questions and tries to determine what the item is.

Solving conundrums is another good way to learn how to eliminate possibilities. Conundrums, also called two-minute mysteries or stories with holes, are short stories that include some sort of secret. A leader reads the story aloud, then the class starts to ask yes-or-no questions to try to determine the mystery. (Students who have heard the story before should not be allowed to play—they'll still enjoy watching their classmates try to solve it.) The leader can also answer that a question is irrelevant and can ask the questioner to be more specific about the question. Here is an example:

> A man lives in a high-rise apartment building. Every day when he goes to work, he gets into the elevator on the 30th floor (where his apartment is) and rides down to the 1st floor and goes out of the building to work. When he comes home in the evening, most of the time he rides the elevator to the 12th floor, gets out there, and walks up the stairs to the 30th floor. Occasionally, however, he rides the elevator all the way from the 1st floor to the 30th floor. Why doesn't he always go straight home?

Possible questions and answers:

Q: Does he visit someone on the 12th floor?

A: No.

Q: Does this have something to do with his job?

A: No.

Q: Is the elevator broken?

A: No.

Q: Is he trying to get some exercise?

A: No.

Q: Are there other people in the elevator?

A: Specify.

Q: Are there other people in the elevator when he gets out on the 12th floor?

A: No.

Q: Are there other people in the elevator when he rides the elevator all the way to the 30th floor?

A: Yes.

Q: Does the mystery have something to do with the other people who may or may not ride the elevator?

A: Yes.

And so on. The answer to this conundrum is that the man isn't tall enough to reach the button for the 30th floor. He can only ride to that floor if someone else is in the elevator to push the button for him.

Here are some other famous conundrums and their answers. Many of these stories tend to be about death, but we tried not to list too many of those here. Conundrums are featured in several books, including in the Cooperative Learning Section of the *Discovering Geometry Teacher's Guide and Answer Key*, third edition (Key Curriculum Press, 2003). You can also make up your own conundrums from unusual newspaper stories.

Story: A man is afraid to go home because a man with a mask is there.
Explanation: The man is on third base in a baseball game, and the man with the mask is the catcher.

Story: A woman pushed her car past the hotel and smiled.
Explanation: The woman is playing Monopoly, and she pushed her game piece (the car) past a hotel on Boardwalk that was not hers.

Story: Inside a windowless room there are three light bulbs. Outside the room there are three light switches. Each of the light switches outside the room controls one of the three bulbs inside the room. You are allowed to enter the room once. At that time you have to be able to figure out which light switch controls which bulb.
Explanation: Turn on one of the light switches for five minutes or so. Then turn it off. Turn on another light switch and enter the room. The switch you just turned on controls the light bulb that is on. The first switch you turned on controls the warm light bulb that is off. The third switch controls the cold light bulb that is off.[1]

Story: A man died with a hole in his suit.
Explanation: The man was a scuba diver (or astronaut), and his suit was punctured.

Story: A man entered a large building, looked around, and then left. A few minutes later a police officer came in and arrested two people.
Explanation: The building was a sports arena for a basketball game. The man's car had been stolen. In the car were two tickets to the basketball game that night. The man showed up at the sports arena and found two people sitting in his seats (wearing his clothes). The police officer came in and arrested the two for car theft.[2]

Story: A woman paid to enter a show, won two awards, and was embarrassed.
Explanation: The woman paid to enter her dog in a dog show. The woman won the award for looking most like her dog, and her dog won for ugliest dog.[3]

[1]Contributed by Sierra College student Erin Pierce.
[2]Contributed by Donna O'Neil from a newspaper story. This really happened.
[3]Contributed by Linda McDonald from a news story.

Story: The firefighter immediately recognized the cause of the fire.
Explanation: At the scene of a grass fire was a dead bird and a singed snake. The bird had been carrying the snake when it fell across two power lines. The bird and the snake were electrocuted, and the snake caught on fire. When the snake and the bird fell to the ground, the grass caught fire.[4]

Story: Thirty people are found dead in a cabin on a hillside.
Explanation: They were found dead in the cabin of an airplane that had crashed on the hillside.

Story: A man and his young son were injured in a car accident. They were taken to the hospital. When the boy was wheeled into the operating room, the surgeon looked at him and said, "I can't operate on this boy; he is my son."
Explanation: The surgeon is the boy's mother. (*Note:* This may not be as puzzling in the twenty-first century as it would have been in the 1950s. But maybe it will be. How far have we come?)

[4]Contributed by Linda McDonald from a news story.

Notes on Text Problems

■ **PENNY'S DIMES, PART 2**

It often helps to determine which clues are the most useful and start with those. For example, if one clue is that a number is even and another is that the number is a multiple of 10, the latter clue is far more useful and better to start with.

In the write-up shown in the text, the students eliminated a lot of possibilities by not writing them down. They wrote down a list of the multiples of 5, thus eliminating numbers between 1 and 100 that are not multiples of 5.

■ **DOWN ON THE FARM**

The students will need to be willing to reread sections of this problem write-up.

■ **DOWNTOWN DELI**

Again, the students need to make sure to write out all possibilities first. A key part of this problem is that an important piece of information (the product) is missing. But it is enough to know that Seymour was able to solve the problem using that information. Knowing that the hidden information helped solve the problem allows students to eliminate any possibility where that information would not be helpful.

©2004 Key College Publishing, *Instructor Resources: Crossing the River with Dogs*, Johnson/Herr/Kysh

Notes on Problem Set A

EGGS IN A BASKET

This is a good Problem of the Day. There are many approaches to this problem. Your students will learn a lot from the presentations.

WOW, WOW, SO COOK!

This is a good Problem of the Day the night before you use the next problem in class.

NELSON + CARSON = REWARD

This is an excellent problem for students to practice a write-up. No guessing is required; all letters can be determined by eliminating possibilities. Have the class solve the problem in groups, which will take 20 to 30 minutes. Walk around the room, occasionally giving hints. A good hint is "concentrate on the second column from the left." How can E + A = E? Let the students figure it out. After they finish solving it, suggest they re-create the order in which they figure out each letter—which will help them do their write-up. Then, for homework, have them write up their solution, explaining how they solved the problem. Grade it with the same rubric that you are going to use to grade problem sets.

THE THREE SQUARES

This is another problem that will require more than one strategy. The students should be starting to accept that (1) some problems will require more than one strategy to be solved efficiently and that (2) it is often a positive skill to start with one strategy and then decide to change to another one. Students should be showing signs of achieving another goal of this course: They should be more persistent in solving problems.

©2004 Key College Publishing, *Instructor Resources: Crossing the River with Dogs*, Johnson/Herr/Kysh

Notes on Problem Set A *(continued)*

■ **TO TELL THE TRUTH**

This is a very difficult problem. It makes an excellent Problem of the Day and should lead to a good discussion. Students need to understand that liars always lie and truth tellers always tell the truth. So, when someone is asked, "Are you a truth teller?" the answer must be "Yes." This is an important step in solving the problem. To get students to see this, have each student choose to be a liar or a truth teller, and then ask them, one by one, "Are you a truth teller?"

However, if Dog 1 does not answer the question, for example, if he says, "The sky is blue," then everything changes. Now Dog 2 is a liar, and more analysis follows. Another issue that should also arise is what it means for the statement "I am the only Joe" to be false. There are many possibilities, including the less obvious such as that there are no Joes or that the speaker is one Joe of many.

All of these issues should be discussed in class.

Good Quiz Problems

For a one-problem quiz on this strategy, consider one of the following problems from Problem Set A, Version 2. We suggest that you grade this using Version 2 of the rubric described on pages 30–31 of this text, but ask the students to make their explanations very brief.

1. Sitting in the Park

2. Dancing in PE

3. Marble Arrangements

4. State Quarters Collection

Questions from the Reading

On a separate piece of paper, answer each question with a complete sentence or sentences. Use the information from your textbook to answer each question. Questions with "explain" or "compare and contrast" need at least a well-formed paragraph and possibly diagrams.

1. What is a general strategy to complete the Twenty Questions game with as few questions as possible?

TEXT BEFORE WHO IS LYING?

2. What two things must be done when seeking contradictions?

3. What can you conclude if you work an assumption through and a contradiction is reached?

4. What can you conclude if you work an assumption through and everything seems to work?

DOWN ON THE FARM

5. How is it established that $E = 1$?

6. Why must the letter O equal 0?

7. What is the purpose of the chart for T, N, and M on page 53?

DOWNTOWN DELI

8. What problem-solving strategies did Richard use?

9. Why did Richard put 45 and 49 on the rows shown in the chart below?

FS #1	FS #2	CB
3	47	45
3	47	49

Questions from the Reading (continued)

10. What shortcut did Richard use in finding the last digit of the product?

11. What did Richard not know that Seymour knew?

12. What is the reasoning behind why Seymour can figure out the number even though Gus can remember only the last digit of the product?

13. After the Downtown Deli problem, what does the text say is a key element in a lot of puzzle problems?

Answers to Questions from the Reading

1. A general strategy is to ask questions that will allow you to split the remaining possibilities roughly in half and eliminate half at a time.

WHO IS LYING?

2. The two things are (1) make an assumption and (2) apply the assumption.
3. You can conclude that the assumption was incorrect.
4. It *only* confirms that an existing possibility is still an existing possibility. Other assumptions are still possible.

DOWN ON THE FARM

5. Siobhan notes that E is a carry digit from the thousands column.
6. The letter O must equal 0 because O + A = A and there is no carrying from the previous column.

7. The purpose is to list all possibilities for these three letters, assume one of them is correct, and then follow it through to seek contradictions.

DOWNTOWN DELI

8. Richard used a systematic list and eliminated possibilities.
9. Richard didn't know on which side of the Fast Stop store the Circle B store would be located.
10. Richard's shortcut was to multiply only the last digits.
11. Richard didn't know the product.
12. Seymour must be able to tell because the last digit of the product is unique to all the remaining possibilities.
13. The text says that even if you don't have all the information, the existence of someone else who has the information means that you can eliminate more possibilities and solve the problem.

Text Problems

©2004 Key College Publishing, *Instructor Resources: Crossing the River with Dogs,* Johnson/Herr/Kysh

PENNY'S DIMES, PART 2

Penny's favorite coin is the dime, as we saw in Chapter 2. Since we last saw Penny, she has spent some of her dimes and has acquired some more. She doesn't know how many she has now, but she knows she has fewer than 100. One day she was arranging them on her desk in different ways. She found that when she put them into piles of 2, there was 1 left over. When she put them into piles of 3, again there was 1 left over. The same thing happened when she put them into piles of 4. She then tried putting them into piles of 5 and found that there were none left over. How many dimes does Penny have? Solve this problem before continuing. (There is more than one correct answer.)

©2004 Key College Publishing, *Instructor Resources: Crossing the River with Dogs,* Johnson/Herr/Kysh

WHO IS LYING?

Jim tells lies on Fridays, Saturdays, and Sundays. He tells the truth on all other days. Freda tells lies on Tuesdays, Wednesdays, and Thursdays. She tells the truth on all other days. If they both say "Yesterday I lied," then what day is it today? Solve this problem before continuing.

©2004 Key College Publishing, *Instructor Resources: Crossing the River with Dogs,* Johnson/Herr/Kysh

DOWN ON THE FARM

Gordon and Pearl lived on a farm in Concordia, Kansas, with their father, Emil, and their mother, Olive. One day Gordon asked his father, "Dad, what happened to that cat I used to have?" Pearl, overhearing this, said, "Yeah, and I used to have a horse. Where is she?" Emil replied, "Gordon's tomcat and Pearl's old nag were not much use. I traded them for my new goat."

Olive then said, "Hey, that sounds like a good cryptarithmetic problem. Let's see if we can solve it." She wrote down GTOM + PNAG = EGOAT. Each letter stands for a different digit, 0 through 9. No two letters stand for the same digit. Determine which digit each letter represents.

There are two clues. $G = 5$, and A represents an odd digit. Do this problem before reading on.

©2004 Key College Publishing, *Instructor Resources: Crossing the River with Dogs,* Johnson/Herr/Kysh

©2004 Key College Publishing, *Instructor Resources: Crossing the River with Dogs,* Johnson/Herr/Kysh

Text Problems *(continued)*

DOWNTOWN DELI

Seymour owns his own business. He makes deli sandwiches, which he wraps to retain their freshness and then distributes to several convenience stores for resale. One of his favorite sandwiches is the Sausage and Meatball Combo, but it has a very low distribution. In fact, only three stores take deliveries of the Sausage and Meatball Combo: two Fast Stop stores and one Circle B store.

One morning, Seymour suffered an unfortunate accident. He slipped on the floor and banged his head. He seemed to be fine, except that when he was out on his delivery route, he couldn't remember which streets the three Sausage and Meatball Combo stores were on. The streets were numbered from 1st Street up to 154th Street, and he remembered that the two Fast Stop stores were on streets whose numbers added up to 50. He also remembered that the Circle B store was two streets away from one of the Fast Stop stores, and he remembered that he called the Sausage and Meatball Combo his "prime" favorite because all three stores were on prime-numbered streets. Unfortunately, the information he remembered wasn't enough to get him to the stores.

He called his friend Gus, and Gus remembered that Seymour had told him the **product** of the numbers of the streets the stores were on, but Gus could remember only the last digit of the product. This proved to be enough for Seymour, who promptly double-parked, whipped out a pencil, made a systematic list, and eliminated possibilities to find the answer.

Now it's your turn to re-create Seymour's heroics.

©2004 Key College Publishing, *Instructor Resources: Crossing the River with Dogs,* Johnson/Herr/Kysh

©2004 Key College Publishing, *Instructor Resources: Crossing the River with Dogs,* Johnson/Herr/Kysh

Problem Set A, Version 2

1. SITTING IN THE PARK

Mary Ann didn't know if her eyes were playing tricks on her or what. When she first looked at the people in the park from the hill she was sitting on, it looked as if they were all in groups of two. Those groups of two seemed to drift together to form groups of four, and then there were two people left over. The next time she looked, they appeared to be in groups of three, but the same two were left over. As she double-checked, it seemed as though those groups of three completely transformed themselves into groups of five, and this time there were four left over. There were fewer than a hundred people at the park. How many were there?

2. DANCING IN PE

Miss Von Thaden tried very hard to make sure that nobody was left out during the PE classes in which her students danced. She thought she had a correct head count, so she told them to pair up. This didn't work because there was one person left out. She then told them to get into groups of five, but this didn't work either because again there was one person left over. So she tried setting up groups of three, but unfortunately there was one left over. Finally, she decided to try groups of four. Again there was one person left out. There are fewer than 80 students in the dance class. How many were present on that particular day?

3. MARBLE ARRANGEMENTS

Garrett was arranging marbles on his bedroom floor one rainy day when he was really bored. He arranged all of his marbles into piles of three, but there were two marbles left over. So he started over again and put all of his marbles into piles of four, but again there were two left over. He started getting frustrated, so he put all of his marbles into piles of five, but again there were two left over. Then he tried piles of six, but again there were two left over. He gave up in complete frustration and put all of his marbles back in his closet. If Garrett has more than 6 but fewer than 100 marbles, how many marbles does he have? What would have happened if Garrett had arranged them into piles of two?

©2004 Key College Publishing, *Instructor Resources: Crossing the River with Dogs*, Johnson/Herr/Kysh

4. STATE QUARTERS COLLECTION

Will was counting the new state quarters that he had been collecting. He started organizing them into piles. He put all of the quarters into piles of three but found that there were two left over. Then he put them into piles of five and found that there were three left over. Then he put them into piles of seven and found that there were five left over. Finally, he put them in piles of two, and there weren't any left over. Will has fewer than 100 quarters. How many quarters does he have?

(*Note:* If you don't understand what "piles of . . ." means, this example may help. Piles of three means that there are three quarters in each pile. Two left over means that after putting all of the quarters in piles of three, there were still two left over that couldn't be made into a pile of three.)

5. A CUBE ROOT

The cube root of 68,921 is an integer. Without using a calculator, determine what that integer is by eliminating possibilities.

Five Cryptarithms

The rules for the Bored Teenagers, One-Hour Casserole, Napa Valley Tour, Shutout, and Secret to Monopoly cryptarithmetic problems are the same as always: Each letter stands for one and only one digit.

©2004 Key College Publishing, *Instructor Resources: Crossing the River with Dogs,* Johnson/Herr/Kysh

Problem Set A, Version 2 (continued)

6. BORED TEENAGERS[5]

A family was taking a trip. While driving through a rural area, the teenagers complained of being hungry and bored. "Look out the window," they were told. "All we see are cows, a barn, and trees," they replied.

"That's a cryptarithmetic problem!" exclaimed their mother, Suzanne. "Cows + Barn = Trees. Why don't you solve it, then we'll stop for a bite to eat."

The teenagers were incredibly unimpressed by the puzzle, but their hunger overcame their objections and they solved it in record time. Unfortunately, they were still in the country and there was no place for them to stop, so they had to suffer a bit more until they reached civilization.

There are three versions of this problem; each one comes with its own hint. The first two versions have the same solution, but you arrive at it a different way for each version. The third version has two different solutions.

Each letter stands for one of the digits 0 to 9, and no two letters stand for the same digit.

COWS + BARN = TREES

Version 1 *Hint:* The digits in COWS are all even and in descending order.

Version 2 *Hint:* The digits in COWS are all even, and S = 2.

Version 3 *Hint:* The digits in COWS are all even, and S does not equal 2.

7. ONE-HOUR CASSEROLE[6]

Marta had to bake a casserole for one hour, but her kitchen timer was broken. Her father suggested that she use the antique hourglass on the curio shelf. He told her to flip it over and after one hour all the sand should be at the bottom of the glass.

Marta realized that this was also a cryptarithmetic problem: Hour + Sand = Glass. She worked on this problem and forgot to look at the hourglass. By the time she finished the problem, the casserole was burned, so they had to go out for dinner.

Each letter stands for one of the digits 0 to 9, and no two letters stand for the same digit.

HOUR + SAND = GLASS   *Hint:* D = 8

[5,6]Sierra College student Suzanne Goodell wrote these problems.

3 • ELIMINATE POSSIBILITIES   **81**

©2004 Key College Publishing, *Instructor Resources: Crossing the River with Dogs,* Johnson/Herr/Kysh

Problem Set A, Version 2 *(continued)*

8. **NAPA VALLEY TOUR**[7]

Carrie and Bruno were visiting the Napa Valley. While driving down the road, they saw a sign that read "Free Tour" with the word "Wines" below. Bruno wanted to take the tour, but Carrie, the designated driver, felt that Bruno had already had as many free tours as he could handle. An argument ensued.

Finally, Carrie agreed that they would take the free tour if Bruno could solve the cryptarithmetic puzzle created by the sign: Free + Tour = Wines. If he could not solve this problem, then she would drive back to the hotel and get him some coffee. To be fair, she gave him one hint: The letter O = 0.

Bruno could not solve the problem, so they drove back to the hotel where he went to the restaurant, ordered Irish coffee, and sang pub songs for the rest of the night.

Could you take the free tour? Each letter stands for one of the digits 0 to 9, and no two letters stand for the same digit.

$$FREE + TOUR = WINES \qquad Hint: O = 0.$$

9. **SHUTOUT**

$$
\begin{array}{r}
S\ O\ C\ C\ E\ R \\
+\ G\ O\ A\ L\ I\ E \\
\hline
N\ O\ S\ C\ O\ R\ E
\end{array}
$$

10. **SECRET TO MONOPOLY**

The victor in a game of Monopoly is often the person who has the most houses and hotels. You might say that the person with the most houses and hotels controls the game. *Hint:* O = 7.

$$
\begin{array}{r}
H\ O\ U\ S\ E\ S \\
+\ H\ O\ T\ E\ L\ S \\
\hline
C\ O\ N\ T\ R\ O\ L
\end{array}
$$

[7]Sierra College student Suzanne Goodell wrote this problem.

Problem Set A, Version 2 *(continued)*

11. FRUITS AND VEGETABLES CRISSCROSS

Fill in the grid below, using each of the given words only once.

FOUR	FIVE	SIX	SEVEN	EIGHT	NINE	TEN	ELEVEN
kiwi	apple	banana	apricot	broccoli	cranberry	blackberry	boysenberry
pear	lemon	orange	kumquat	cucumber	greenbean	strawberry	
plum	peach	potato	pumpkin	lima bean	raspberry		
		tomato		zucchini	tangerine		

12. **BASEBALL CRISSCROSS**

Fill in the grid below, using each of the given words only once.

THREE	FOUR	FIVE	SIX	SEVEN	EIGHT	NINE
bat	balk	bases	batter	bull pen	ballpark	sacrifice
ERA	ball	error	corner	catcher	baseball	
fly	bunt	homer	double	fielder	grounder	
hit	foul	pitch	relief	manager	hummbaby	
low	runs	pop-up	single	shutout	pitchout	
one	safe	score	strike	squeeze		
out		steal	triple			
RBI			umpire			
two						

TEACHING RESOURCES

©2004 Key College Publishing, *Instructor Resources: Crossing the River with Dogs,* Johnson/Herr/Kysh

Problem Set B, Version 2

1. LOG CUTTING

Selena cut a log into three pieces in 6 minutes. She then cut a similar log into five pieces. How long should the job have taken if she worked at the same rate she had worked when cutting the first log?

2. WEIRD WORDS

Write three-letter words according to this plan: The first letter must be *b* or *c*. The second letter must be a vowel. The third letter must be chosen from *t*, *v*, or *x*. How many such words can be written? (These do not have to be English words, or real words of any other language for that matter.)

3. GETTING AROUND TOWN

The state capitol building in California is located in Sacramento. It's beautiful, but getting there is sometimes very tough. I actually needed to get to the corner of 12th and K streets from my location at the corner of 13th and N streets. Unfortunately, due to all the one-way streets, it was going to be very difficult. Sacramento has numbered streets that run north and south, and lettered streets running east and west. The numbers get larger as you drive east, and the letters get later in the alphabet as you drive south. The one-way and two-way streets are as follows:

Two-way, north and south: 6th, 11th, 13th, 14th, 17th

Two-way, east and west: M, O

One-way, north to south: 7th, 9th, 12th, 15th

One-way, south to north: 8th, 10th, 16th

One-way, west to east: J, N

One-way, east to west: I, L, P

K Street is a pedestrian shopping mall with no cars allowed. (Cars can cross K Street on numbered streets, but no cars can drive on K Street.)

Now, this wouldn't be complicated except for the capitol building. It takes up ten square blocks, with no streets going through it from 10th to 15th streets and L to N streets. (M Street, 11th, 12th, 13th, and 14th streets are blocked by the capitol. L, N, 10th, and 15th all are continuous along the edges of Capitol Park.) And to top it all off, O Street doesn't exist between 12th and 13th streets for some reason. How could I drive from 13th and N to 12th and K?

©2004 Key College Publishing, *Instructor Resources: Crossing the River with Dogs*, Johnson/Herr/Kysh

Problem Set B, Version 2 *(continued)*

4. SLEEPY BABIES

The Simpson family gave birth to twins recently. In the twins' early months, they didn't do much except sleep and eat and sleep and smile and sleep. In fact, even when they were awake they did a lot of yawning. The twins' aunt was visiting one day when the father commented about their continuous yawns: BABY + BABY = YAWNS. The aunt said, "That sounds like a good problem."

```
    B  A  B  Y
 +  B  A  B  Y
 ─────────────
 Y  A  W  N  S
```

Each letter represents one of the digits from 0 to 9. No two letters can stand for the same digit. What sum does this problem represent?

5. VIDEO GAMES

My brother was playing a goofy video game at the arcade. He got 2 points for destroying a tank, 3 points for destroying a helicopter, 6 points for destroying a jet, and an extra point if the jet he blasted was equipped with atomic weapons. He scored 18 points before his quarter ran out. In how many different ways could he have reached that score?

TEACHING RESOURCES

©2004 Key College Publishing, *Instructor Resources: Crossing the River with Dogs*, Johnson/Herr/Kysh

Problem Set B, Version 3

1. **THE BIKE PATH AROUND THE LAWN**

We have a 27-by-20-foot lawn. We want to pour concrete for a sidewalk 3 feet wide around the lawn. To make the mold for the concrete, we will need to buy some 2-by-4-inch lumber. How many feet of 2-by-4-inch lumber will we need just for the perimeter of the walk? (Consider both the inside and outside perimeters.)

2. **FAST FOOD**

Patty went to a hamburger joint. She considered the choices below. List all of the possibilities for Patty if she has one meat, a side dish, and a drink.

Meat: Hamburger, chicken nuggets, hot dog

Side dish: Fries, onion rings

Drink: Coffee, milk, soda, milkshake

3. **A DARING CATFISH**

Lindsey, a daring catfish, went on a journey. She left her home early one morning and decided to explore the upper part of the river. Each day she swam about 5 miles upstream against the current, but as she tired and rested, she slipped back with the current about 2 miles each night. Eventually, she went just under 20 miles upstream to a fork in the river and then swam downstream in the other branch of the river. Swimming downstream was easier, since she was able to swim at her regular speed and be carried along with the current. Then at night when she rested, the current carried her in the direction she wanted to go. She went 35 miles downstream from the fork in the river and finally reached a lake. How long from the day she left home did it take her to get to the lake? (Assume that night and day are each 12 hours long. Also assume that the current in both branches of the river is the same strength.)

©2004 Key College Publishing, *Instructor Resources: Crossing the River with Dogs*, Johnson/Herr/Kysh

Problem Set B, Version 3 *(continued)*

4. FIND MY NUMBERS

From the clues below, figure out what five numbers I am thinking of.

1. All the numbers are odd.

2. All the numbers are two-digit numbers.

3. The numbers add up to 97.

4. None of the numbers are the same.

5. The largest number is 25.

6. The second-largest number is 4 more than the third-largest number.

5. FOOTBALL SCORES

In how many different ways can the Denver Broncos score 21 points in a football game? Points are scored as follows: A safety scores 2 points, a field goal 3 points, a touchdown 6 points. After a touchdown is scored, the team may get 1 extra point for a kick, 2 extra points for a 2-point conversion, or 0 points if they fail on either of these.

TEACHING RESOURCES

©2004 Key College Publishing, *Instructor Resources: Crossing the River with Dogs,* Johnson/Herr/Kysh

Problem Set B, Version 4: Camping

1. TRAILERS

I was looking around the various campsites on a recent camping trip and noticed that all the trailers present had these characteristics:

Stripes were green, brown, yellow, or blue.

Windows were sliding or jalousie.

Length was 18 ft, 24 ft, or 32 ft.

Each trailer in the campsite was different in some way from every other trailer. Every possible different trailer was present. How many trailers were there?

2. TRAILER PAD

Each campsite featured a trailer driveway. The driveway measured 21 feet across by 42 feet back and was surrounded on three sides by a concrete curb 6 inches wide and 8 inches tall. There was no curb on the front of the driveway so that a trailer could be driven in. What was the volume of the concrete in the curb?

3. KITE CHASING

I was flying a kite during the camping trip, and my dumb dog was chasing the kite's shadow on the ground. He started out right next to me. He then took three steps ahead, two steps back, three side steps to his right, and two side steps to his left. He executed this four-move sequence four times. Then he turned 180 degrees around. This time his sequence was three steps ahead, two back, two side steps to his right, and three side steps to his left. He also executed this sequence four times. Where did he end up?

©2004 Key College Publishing, *Instructor Resources: Crossing the River with Dogs*, Johnson/Herr/Kysh

Problem Set B, Version 4: Camping *(continued)*

4. **FIELDING PRACTICE**

I played baseball with my son on the camping trip. We invented a game called Fielding Practice. He got 10 points for catching a pop fly and making a good throw, 8 points for catching a pop fly and making a bad throw, 7 points for fielding a grounder and making a good throw, 5 points for fielding a grounder and making a bad throw, and 1 point for a good throw after making a catching error (on either a pop fly or a grounder). He scored 20 points in this game. In how many ways could he have scored 20 points?

5. **SLEEPING IN THE TENT**

After dinner we sat around trying to make up word-arithmetic problems. My wife came up with this one. She told me that the sum of the unused digits is 16. Find the solution.

$$
\begin{array}{r}
T\ E\ N\ T \\
+\ S\ I\ T\ E \\
\hline
S\ L\ E\ E\ P
\end{array}
$$

4 Use Matrix Logic

Like the strategies covered in Chapters 2 and 3—making a systematic list and eliminating possibilities—matrix logic is a strategy with its roots in organizing information. It is also an extension of the eliminating possibilities strategy. In matrix logic problems, though, a specific answer or set of specific answers is to be uncovered.

This strategy is a paper form of the game Clue. Students must work slowly and carefully, because estimation skills are thrown out the window in matrix logic. An error in matrix logic will not necessarily show up as an unreasonable answer at the end. This strategy is also extremely difficult to write about because a number of steps and complex logical connections are involved. For this reason, we advocate a notation that indicates from which clue each negation originates. This notation can help students check their work, explain their reasoning processes, and work together.

Matrix logic is typically used when setting up students' schedules in high schools. It is also often used, though not as uniformly, in setting up teachers' schedules. It can be used for distributing tasks to various people, some of whom can perform several tasks and others who are more specialized.

Note on a Text Problem

 **COAST TO COAST**

This problem hinges on using the methods of indirect proof. In the text, this method is referred to as *seeking contradictions*.

Notes on Problem Set A

Problems 1–7 are one-chart problems, in increasing order of difficulty, involving four or five people. Problem 8, Volleyball Team, is a three-chart problem involving three people, which makes a good introduction to three-chart problems. If students are working on several of these problems in groups, the instructor can walk around to each group and briefly explain the setup for a three-chart problem as groups get to Problem 8.

Problems 9–11 are three-chart and four-people problems. Problem 12, Anniversaries, is a six-chart and three-people problem. Problem 13, Paying the Bills, requires an assumption to be made.

Problem 14, Class Schedules, makes a good Problem of the Day. We suggest that each student present only one person's schedule.

Good Quiz Problems

For a one-problem quiz on this strategy, consider one of the following problems from Problem Set A, Version 2. We suggest that you grade this using Version 2 of the rubric described on pages 30–31, but ask the students to make their explanations very brief. Perhaps have the students explain just one or two of the conclusions they drew from the clues.

1. The Hobbyists

2. Mixed Doubles Tennis

3. Summer Jobs

4. Sports Are All Relative

Text Problems

■ FAVORITE SPORTS

Ted, Ken, Allyson, and Janie (two married couples) each have a favorite sport: running, swimming, biking, and golf. Given the following clues, determine who likes which sport.

1. Ted hates golf. He agrees with Mark Twain that golf is nothing but a good walk spoiled.

2. Ken wouldn't run around the block if he didn't have to, and neither would his wife.

3. Each woman's favorite sport is featured in a triathlon.

4. Allyson bought her husband a new bike for his birthday to use in his favorite sport.

Use the matrix on page 69 of the text to work this problem before continuing.

■ OUTDOOR BARBECUE

Tom, John, Fred, and Bill are friends whose occupations are (in no particular order) nurse, secretary, teacher, and pilot. They attended a picnic recently, and each one brought his favorite meat (hamburger, chicken, steak, and hot dogs) to barbecue. From the clues below, determine each man's occupation and favorite meat.

1. Tom is neither the nurse nor the teacher.

2. Fred and the pilot play in a jazz band together.

3. The burger lover and the teacher are not musically inclined.

4. Tom brought hot dogs.

5. Bill sat next to the burger fan and across from the steak lover.

6. The secretary does not play an instrument or sing.

Use the chart on page 75 of the text to work this problem before continuing.

Text Problems *(continued)*

COAST TO COAST

Four women live in different cities. One of the cities is San Francisco. Determine which city each woman lives in.

1. The woman from Charleston (South Carolina), the woman from Gainesville (Florida), and Riana are not related.

2. Wendy and the woman from Provo are cousins.

3. Neither Phyllis nor Wendy is from the West Coast.

4. Ann is from a coastal city.

Solve this problem before reading on.

©2004 Key College Publishing, *Instructor Resources: Crossing the River with Dogs,* Johnson/Herr/Kysh

©2004 Key College Publishing, *Instructor Resources: Crossing the River with Dogs,* Johnson/Herr/Kysh

Problem Set A, Version 2

1. THE HOBBYISTS

Elaine, Leisa, Brittney, and Consuelo each have a hobby: model railroading, building model airplanes, rocketry, or raising tropical fish. Match each woman to her hobby by using the clues below.

1. Leisa has never met the person who does rocketry.

2. Elaine is a pilot and, ironically, has a hobby that has nothing to do with aeronautics.

3. The rocketry hobbyist, the railroader, and Brittney are friends.

4. Leisa's hobby involves public transportation.

2. MIXED DOUBLES TENNIS

Amaya, Ostergard, Blue Cloud, and Katricz are the last names of Timothy, Diana, Mack, and Sherry. They are all playing in a mixed doubles tennis tournament. Two people are on each team. There is one man and one woman on each team. Determine the full name of each player by using the clues below.

1. Mack is a better player than Ostergard.

2. Timothy is Diana's partner.

3. Sherry and Katricz are on the same team.

4. Amaya is known for his wicked serve.

5. Katricz is an opponent of Ostergard.

6. Blue Cloud is an opponent of Amaya.

3. SUMMER JOBS

Four friends—LaTisha, Zack, Steve, and Michelle—are working in summer jobs. The jobs they have found this summer include food server, lifeguard, construction worker, and clerk at a grocery store. Determine who is working which job by using the clues below.

1. The person doing the food-serving job really likes his work.

2. Zack and the lifeguard have known each other for years.

3. Both Michelle and the lifeguard are outside most of the time, and the other two are inside most of the time.

4. LaTisha and the person working construction met on their job last summer.

5. Neither Zack nor the food server worked last summer.

Problem Set A, Version 2 (continued)

4. SPORTS ARE ALL RELATIVE

Determine each person's favorite sport. (One person's name was Michael, and one sport was tennis.)

1. Two siblings had the same favorite sport.

2. Stefan liked a sport everyone else hated.

3. Ryan did not like bowling.

4. Bonnie was one of the two who liked basketball.

5. Ryan is Bonnie's cousin.

5. MATH DEPARTMENT MEETING

JAMES: I've gathered the five of you here to discuss the class schedule for next semester. There are five math classes left to staff: statistics, calculus, finite math, algebra, and technical math.

FARA: I've been teaching algebra a lot recently. I want a break.

JAMES: Okay. Cliff, you've been teaching technical math every semester. Do you want it again?

CLIFF: Yeah, I'll take it.

JAMES: Okay. What about finite? Who's interested?

LAURIE: I am.

FARA: So am I.

(No one else said anything.)

JAMES: Well, okay. Let's talk about calculus. Who is interested?

FARA: I am.

(Elaine and Maile said they were too.)

JAMES: Can we agree on stats?

MAILE: I don't want it.

LAURIE: I've been doing it, but I want a break.

JAMES: Okay, now how about algebra?

ELAINE: I'd like to continue teaching algebra. Does anyone else want it?

JAMES: (after waiting a few seconds) You've got it, Elaine.

(James made some notes and showed them to Fara.)

JAMES: Is this okay, Fara? You've never taught it before . . .

FARA: Yeah, I'll give it a try.

Which class did each person end up with?

TEACHING RESOURCES

©2004 Key College Publishing, *Instructor Resources: Crossing the River with Dogs*, Johnson/Herr/Kysh

Problem Set A, Version 2 (continued)

6. NEXT YEAR AT COLLEGE

Akinte, Chuck, Jenny, and Norma are all going to college next year. Use these clues to find their full names and their respective colleges.

1. Akinte is not going to the University of Memphis or to Penn Valley, and the same is true of Van Hee.

2. Chuck applied to Penn Valley but decided to go elsewhere.

3. Penigar and the student going to Penn Valley are avid tennis players.

4. Norma, the person going to St. Mary's, and Penigar all want to major in biology.

5. Both Norma and Reynoso considered Penn Valley as their second choice.

6. Akinte and Oslowski both were summer counselors at Sierra Tech, even though neither one will go there during the regular year.

7. LOOMIS DAY PARADE

We didn't know who to root for to win Best of Parade in the Loomis Day Parade. We knew about half the people in the parade. We saw Stacy, Mort, Wayne, and Cloe, each participating in a different group. In fact, among them they finished first through fourth place as Best of Parade, though not necessarily in that order. Match each name with the group or activity and the place each group was awarded.

1. Stacy's group placed higher than the dance company.

2. One of the band members had a bet with Wayne, but Wayne's group came in higher.

3. Cloe and Mort both used to belong to the baton twirlers' group but were too busy with their other activities to continue.

4. Mort's group placed higher than the horseback riders but lower than the band.

5. Cloe, the trumpeter, and a member of the first-place group all go to Del Oro High School.

©2004 Key College Publishing, *Instructor Resources: Crossing the River with Dogs*, Johnson/Herr/Kysh

Problem Set A, Version 2 *(continued)*

8. STUDENT ACHIEVEMENT AWARDS

The top students (Velma, Matt, Ginny, and Clyde) in four courses (English, computers, physics, and mathematics) at Des Moines High were honored at a pre-graduation ceremony. Match each student's first name, last name, and course.

1. The girl who won in computers was also a runner-up in English.

2. Sholseth moved to Des Moines eight months ago and couldn't get into the computer class because it was full.

3. Both Sholseth and Clyde were in the same math class with the top math student.

4. The top English student, the one named Kinsella, and Ginny have lived in Des Moines all their lives.

5. The boy who won the honors in math barely beat out Macomber. Both boys were runners-up in computers.

6. Perata knew she was close for math honors.

9. NOVEMBER ELECTIONS

Four political allies are running for different offices in the November elections: president, governor, assembly member, and senator. Their first names are Margurite, Darcie, Brent, and Amir. Their last names are Cusack, Tomfohrde, Wyckoff, and Hardy. Match each person's full name to the office she or he is running for by using the clues below.

1. Of Margurite and Tomfohrde, one is running for senator and the other is running for governor, though not necessarily in that order. Candidate Cusack did not enter either race.

2. Darcie has been helping the candidate for president campaign in her area.

3. Brent is not running for president and neither is Hardy.

4. Wyckoff and Darcie once opposed each other in a city council primary race.

5. The candidate for governor is neither Darcie nor Tomfohrde.

6. Margurite has never run for office before.

TEACHING RESOURCES

©2004 Key College Publishing, *Instructor Resources: Crossing the River with Dogs*, Johnson/Herr/Kysh

Problem Set B, Version 2

1. PACIFIC RIM

Lois Onassis wishes to establish shipping routes around the Pacific Rim. She wants to set up freight services in the following eight ports: Los Angeles (United States), Anchorage (United States), Tokyo (Japan), Taipei (Taiwan), Manila (Philippines), Auckland (New Zealand), Valparaíso (Chile), and Lima (Peru). She wants a separate shipping route connecting each pair of ports. How many such routes does she need to establish?

2. PRIME JACKS

Five friends—Amy, Betty, Clarrise, Dawn, and Ellen—each have a certain number of jacks. Amy has the fewest, Betty the next fewest, and so on up to Ellen, who has the most. Each girl has a different number of jacks, and each number is a two-digit prime number ending in 9. The problem with having a prime number of jacks is that they cannot be arranged in piles with the same number of jacks in each pile. For example, if a girl had seven jacks, she could only put them into one pile of seven or into seven piles of one. However, if she had eight jacks, she could arrange them in one pile of eight, two piles of four, four piles of two, or eight piles of one. For this reason, each of the girls occasionally likes to combine her jacks with another girl's jacks to be able to separate them into piles with the same number of jacks in each pile. Which pair of girls can combine their jacks and be able to separate them into same-sized piles in the greatest number of different ways? Which pair combines for the least number of ways?

3. RELATIONSHIPS

Brandon is Rodd's father. Jesse is Rodd's only brother. Tina is Rodd's sister-in-law. Rodd is not married. Mela is Tina's mother-in-law. Jordan is Jesse's son. What relation is Jordan to Mela? What relation is Tina to Brandon?

©2004 Key College Publishing, *Instructor Resources: Crossing the River with Dogs*, Johnson/Herr/Kysh

Problem Set B, Version 2 *(continued)*

4. TO TELL THE TRUTH

Four friends get together. One tells the truth all the time. One lies all the time. One tells the truth on odd-numbered days and lies on even-numbered days. One tells the truth on even-numbered days and lies on odd-numbered days. One day in May, they made the following statements:

Abe: I lied yesterday.

Blanca: Today is the twelfth.

Carol: Yesterday's date was even.

Doug: Carol's statement is true.

Which of the four friends tells the truth on even-numbered days and lies on odd-numbered days?

5. THE ACTION NEWS TEAM

The news team at WOWM in Chicago is an outstanding group of individuals who have banded together to form an excellent broadcasting team. Four of the *most* outstanding are the anchor, the sports reporter, the director, and the producer. Their names, in no particular order, are Alex, Chris, Pat, and Sam. The lengths of time they have been at the station are 5, 10, 15, and 20 years. From the clues below, determine each person's name, job, and years of service to the station.

1. Among the four are two women and two men. One man and one woman are on camera.

2. Alex (short for Alexandra) used to be on camera, but 12 years ago she switched to the off-camera job she currently holds.

3. Chris (short for Christopher) has been at the station longer than the anchor but not as long as Sam.

4. Pat (short for Patrick) has never worked behind the scenes.

5. Sam (short for Samantha) started at the station 10 years before the anchor. Her first job at WOWM was as a lighting technician.

6. The director has never held any other job.

©2004 Key College Publishing, *Instructor Resources: Crossing the River with Dogs,* Johnson/Herr/Kysh

Problem Set B, Version 3

1. DOOR IN AND DOOR OUT

An auditorium has 11 doors numbered 1 through 11. In how many ways is it possible to enter the auditorium through an odd-numbered door and leave through an even-numbered door?

2. JERRY'S AGE

Jerry got a phone call from a telephone solicitor who was taking a survey. First the solicitor asked how old Jerry was. Jerry gave the following clues:

"My age is between 30 and 69, inclusive."

The solicitor asked if he was in his 30's.

"If I am not in my 30's, then my age is a multiple of 4."

The solicitor asked if he was in his 40's.

"If I am not in my 40's, then my age is a multiple of 7."

The solicitor asked if he was in his 50's.

"If I am not in my 50's, then my age is not a multiple of 5."

The solicitor finally asked if he was in his 60's.

"If I am not in my 60's, then my age is not a multiple of 8."

How old is Jerry?

3. KLINGONS LIKE WORD ARITHMETIC TOO

When Captain Picard first visited the Klingon home world with Lieutenant Worf, they visited the home of Worf's brother, Kurn. Kurn was quite fond of word-arithmetic problems similar to the word-arithmetic problems Picard had done on Earth as a boy. Worf challenged Picard to a word-arithmetic problem provided by Kurn. As is the custom for Earth problems, each letter stands for one of the digits 0 through 9, and no two letters stand for the same digit. You may not recognize the words—they are Klingon. Find the sum represented by this problem:

```
    C  R  E  K  L  G
 +  R  U  T  A  N  G
 ─────────────────────
 C  K  G  L  E  L  K
```

4. **THE SHADOW KNOWS**

John Henry was camping in a large, sparsely wooded area in southern Texas. One day he went for a walk. He packed some food and left at 8:00 a.m. on a cloudless, hot day. He walked for 2 miles with his shadow on his left. Then he walked for 5 miles with his shadow in front of him. Then he walked 3 miles with his shadow on his right. Then he walked 1 mile with his shadow behind him. By this time it was 11:00 a.m. He stopped in a meadow and had lunch. Then he fell asleep because he was tired from walking. When he woke up, it was 1:00 p.m. He was a little disoriented from his nap as he set out to walk home. He figured he could reverse his previous distances and shadows and walk back to his camp. So he walked 1 mile with his shadow in front of him. Then he walked 3 miles with his shadow on his left. Then he walked 5 miles with his shadow behind him. Then he walked 2 miles with his shadow on his right. Unfortunately, he didn't arrive back at his camp. Give directions for him to get back to his camp by the shortest route.

5. **THE NEW FALL SEASON**

The program directors for WBC got together to discuss programming for the upcoming fall season. The result was four new shows: a comedy, a drama, a variety show, and a news program. The names of the shows were *Wall Street Blues, All in Favor, Murphy's Law,* and *Fifty-Fifty.* The shows were scheduled to air on Monday, Tuesday, Wednesday, and Thursday. From the clues below, determine the name, day, and type of each show.

1. Tuesday's show (which was not *Murphy's Law*) and the news program had been tried out during the previous spring.

2. *Fifty-Fifty* was not the variety show.

3. The drama (which was not *Fifty-Fifty*) and Monday's show (which was not *Wall Street Blues*) were produced by the same company.

4. *All in Favor* (which was not the drama) and Monday's show (which was not the news program) both featured a person named Bob.

5. Neither *Wall Street Blues* nor the drama would air on Thursday.

Problem Set B, Version 4

1. LUCKY SVEN

I went shopping for a used car a few days ago, and stopped at Lucky Sven's Autorama. One car I liked was priced at $3,211, another car at $2,311, and another at $4,111. I noticed that if you added all of the digits of each price, they added to 7. I talked to the owner, Sven, and he told me that every car was priced like that; the sum of the digits in the price was 7. Every car had a different price, and he never used a zero in the price. None of the cars cost more than $15,000, and none cost less than $1,000. How many different prices did Sven have available to him?

2. LETTER PUZZLE

In the puzzle below, each letter stands for a different digit from 0 to 9. The same letter stands for the same digit throughout the problem. What digit does each letter represent?

$$Y + W = D \qquad D - G = R \qquad W/Y = G \qquad C + M = C$$

$$H \times A = H \qquad Y + Y = W \qquad Y - G = A \qquad H \times G = L$$

3. TWO GUARDIANS[1]

There are two guardians guarding two doors. One of the doors leads to certain death, and the other doesn't. One guardian always tells the truth, and the other guardian always lies. But you do not know which guardian is which. Each guardian makes a statement:

Guardian 1: The other guardian always lies.

Guardian 2: The other guardian would say door number two leads to certain death.

Which door should you take?

4. RELATIVES

Denise had once been married to Ron. They had a daughter named Jeannine. Denise then married Scott, and they had a son named Phillip. Scott had previously been married to Janelle. They had a son named Tom. Ron then married Andrea. They had a daughter named Nancy. Andrea had previously been married to Bo. They had a son named Woody. Some of the kids are half brothers or half sisters, and others are stepbrothers or stepsisters. Other kids are not related at all. Determine all of the relationships among the kids.

---

[1]This problem was written by a student.

Problem Set B, Version 4 (continued)

WAYNE'S WORLD

Four college friends—Wayne, Garth, Dana, and Mike (whose last names are Campbell, Algar, Carvey, and Myers)—love to watch the movie *Wayne's World*. They each play one instrument in a four-piece rock band (guitar, bass, keyboard, and drums). From the clues below, determine each person's full name and what instrument each plays.

1. The friends think it very unusual that the two characters in the movie *Wayne's World* are named Wayne Campbell and Garth Algar, played by actors Mike Myers and Dana Carvey. However, none of the four friends has any of those full names.

2. Wayne and Carvey are seniors. Only one of them sings in the band.

3. Two of the three seniors and the keyboard player (who is a junior) are taking chemistry together. Mike (the other senior) is taking physics instead of chemistry.

4. Only two of the band members sing. One of them is a junior, and the other is Myers.

5. The drummer phoned Algar to get the chemistry assignment and was surprised to find that Algar was not taking chemistry.

6. Neither the guitar player nor the drummer sings.

TEACHING RESOURCES

©2004 Key College Publishing, *Instructor Resources: Crossing the River with Dogs,* Johnson/Herr/Kysh

5 Look for a Pattern

Looking for patterns is a central learning skill of mathematics. But it is not *only* a learning skill; it is also the central theme of mathematics study. Math is often called the study of patterns. It is essential to see and interpret patterns to enjoy the beauty of mathematics. This chapter looks at arithmetic, exponential, Fibonacci, and other types of patterns. It also develops the skill of looking for patterns as a problem-solving strategy. Additionally, looking for patterns will be an integral part of the strategy described in Chapter 9: Solve an Easier Related Problem.

Looking for a pattern is another strategy based on organizing information. By organizing information, you can extract more out of it.

Students must take time to discuss patterns that may seem trivial to you. Many of these patterns are not trivial for students. And, as usual, it is important for students to write out their work, showing their thought processes.

Looking for patterns is an essential part of expanding thinking. Any type of research is based on discovering and describing patterns. In virtually any occupation, people are called upon to notice patterns. Auto mechanics may notice that certain makes of automobiles have trouble with injection systems. Or perhaps they notice early buildup of carbon inside cylinders on certain engines.

Real-life patterns don't have to be occupationally based. For example, you might notice that someone keeps looking at you. That pattern may indicate a romantic interest or, possibly, that you have twigs stuck in your hair. Effective parenting requires the ability to notice patterns—that certain rewards work well with children and that certain stimuli exact negative behaviors from children. In softball, a batter who swings at pitches too far inside may find that the pitcher noticed this pattern and intends to keep on pitching there.

©2004 Key College Publishing, *Instructor Resources: Crossing the River with Dogs*, Johnson/Herr/Kysh

Notes on Problem Set A

More Sequence Patterns makes an easy Problem of the Day. Have students present only one part. Air Show is excellent as a Problem of the Day, as there are many ways to solve the problem. Bees makes a good Problem of the Day because it introduces the Fibonacci sequence.

Good Quiz Problems

We recommend no quiz on this chapter.

Text Problems

SEQUENCES

Find the pattern and predict the next four terms. Then write a sentence that explains your pattern. Solve each problem before continuing.

A. 1, 2, 4, _____, _____, _____, _____,

B. 1, 3, 5, 7, _____, _____, _____, _____

C. 1, 6, 11, 16, _____, _____, _____, _____

D. 1, 4, 9, 16, _____, _____, _____, _____

E. 1, 3, 6, 10, _____, _____, _____, _____

F. 3, 6, 5, 10, 9, 18, 17, 34, _____, _____, _____, _____

G. 1, 3, 4, 7, 11, 18, 29, _____, _____, _____, _____

H. 2, 3, 5, 9, 17, 33, _____, _____, _____, _____

I. 77, 49, 36, 18, _____ (This sequence ends here.)

DODGER STADIUM

Radio broadcasters joke about the number of people who start leaving Dodger Stadium during the seventh inning of baseball games. One evening, during a particularly boring baseball game in which the Dodgers were trailing by six runs after six innings, the fans began to leave at a record pace. After the first out in the top of the seventh inning, 100 fans left. After the second out, 150 fans left. After the third out, 200 fans left. The pattern continued in this way, with 50 more fans leaving after each out than had left after the previous out. The ridiculous thing was, the Dodgers tied the game in the bottom of the ninth inning, and people still kept leaving early. The game lasted ten innings (the Dodgers lost anyway), and the pattern continued through the bottom of the tenth inning. How many fans left early? Work this problem before continuing.

Text Problems (continued)

PITTER PATTER RABBITS' FEET

Tessa wanted to buy a rabbit. She had liked the Easter bunny when she was a kid, so she decided to raise some bunnies of her own. She went to the store with the intention of buying one rabbit, but she ended up with two newborn rabbits, a male and a female. She named them Patrick and Susan. Well, rabbits being what they are (rabbits), it is fairly impossible to have just two rabbits for an extended period of time. She bought them on April 1, 2003. On June 1, she noticed that Patrick and Susan were the proud parents of two newborn rabbits, again one male and one female. She named these new arrivals Thomas and Ursula.

On July 1, Patrick and Susan again gave birth to a male and a female rabbit. She named these Vida and Wanda.

On August 1, Patrick and Susan again gave birth to a male and a female. But Tessa was really surprised to see that Thomas and Ursula also gave birth to a male and a female. Tessa was running out of names, so she didn't bother giving them any.

On September 1, Patrick and Susan gave birth to a male and a female, and so did Thomas and Ursula, and so did Vida and Wanda.[1]

Tessa noticed a pattern to the breeding. A pair of rabbits was born. Two months later they bred a pair of rabbits and continued to breed a pair of rabbits every month after that. Tessa wondered, "If this keeps up, how many rabbits am I going to have on April 1, 2004?"

Do this problem before continuing.

©2004 Key College Publishing, *Instructor Resources: Crossing the River with Dogs*, Johnson/Herr/Kysh

NIGHT OF THE HOWLING DOGS

Shawna liked to jog in the late afternoon. One day she noticed an unusual phenomenon. As she jogged, dogs would hear her and bark. After the first dog had barked for about 15 seconds, two other dogs would join in and bark. In about another 15 seconds, it seemed that each barking dog would "inspire" two more dogs to start barking. Of course, long after Shawna passed the first dog, it continued to bark, as dogs are inclined to do. After about 3 minutes, how many dogs were barking (as a result of Shawna's passing the first dog)? Work this problem before continuing.

©2004 Key College Publishing, *Instructor Resources: Crossing the River with Dogs*, Johnson/Herr/Kysh

---

[1] Actually, Vida was no longer Vida, and Thomas was no longer Thomas. Tessa was worried about maintaining a diverse genetic pool among her bunnies, so she traded the original Thomas and Vida to other breeders and named their replacements with the same names.

TEACHING RESOURCES

©2004 Key College Publishing, *Instructor Resources: Crossing the River with Dogs*, Johnson/Herr/Kysh

Text Problems *(continued)*

 **MILK LOVERS**

Alysia and Melissa and Dante and Melody loved milk. They convinced their older brother, Mark, who did all the shopping, to buy each of them a gallon of milk because they liked it so much. They all put their names on their gallons. One day, they were all really thirsty, and each took ten drinks according to a different system.

Alysia started by drinking half of the milk in her container. Then she drank one-third of what was left. Then she drank one-fourth of what was left, then one-fifth, and so on.

Melissa started by drinking one-eleventh of her milk, then one-tenth of what was left, then one-ninth of what was left, and so on.

Dante started by drinking one-half of his milk, then two-thirds of what was left, then three-fourths of what was left, then four-fifths, and so on.

Melody started by drinking one-half of her milk, then one-half of what was left, then one-half of what was left, and so on.

After each had taken ten drinks, how much milk remained in each container? Work this problem before continuing.

Problem Set A, Version 2

1. SEQUENCE PATTERNS AGAIN

Write the next three numbers in each sequence and explain your pattern.

a. 5, 8, 11, 14, 17, _____, _____, _____

b. −1, 1, 3, 5, 7, _____, _____, _____

c. 2, 3, 5, 8, 12, _____, _____, _____

d. 3, 2, 5, 7, 12, _____, _____, _____

e. −1, 5, 4, 9, 13, _____, _____, _____

f. 4, 5, 7, 10, 10, 15, 13, _____, _____, _____

g. −2, −1, 1, 4, 8, _____, _____, _____

h. 5, 7, 11, 19, 35, _____, _____, _____

2. THE GREAT SALE

There was a great clearance sale going on at Tucker's Department Store. Starting at 9:00 a.m., Asa called his friends Maggie, Spencer, and Pam to tell them about it. Each of Asa's friends called three of their friends in the next half hour but did not call any more friends after those three. Each subsequent person who was informed about the sale then called three more during the next half hour. The pattern of friends calling friends continued until just before 3:00 that afternoon. How many people heard about the sale as a result of Asa and his friends?

3. MMM, MACARONI AND CHEESE

Orie was too excited about his new toy to sit still. He came to the lunch table, ate ⅙ of his macaroni and cheese, then left to play with his new toy, came back, ate ⅐ of what was left, ran and played, came back, ate ⅛ of what was left, and so on. After ten episodes of eating then playing, how much of his original macaroni and cheese was left?

4. BIRD AND BEEF

Paloma and Chuck opened a "bird and beef" stand that featured hamburgers and gourmet squab fillets. On the first day nobody came. On the second day, however, there were two customers. On the third day there were four customers, and on the fourth day there were six customers. After 50 days, how many customers total will have been served if this pattern keeps up? (Note that after 3 days, they had served six total customers, because $0 + 2 + 4 = 6$.)

TEACHING RESOURCES

©2004 Key College Publishing, *Instructor Resources: Crossing the River with Dogs,* Johnson/Herr/Kysh

Problem Set A, Version 2 *(continued)*

5.   MAILING LISTS

Millie was always buying things through the mail. Well, at least *sometimes* Millie bought things through the mail. Okay, actually, *once* she bought something through the mail from the *Fishing in the West* catalog as a present for her granddaughter. The order arrived in January, and so did five other catalogs, apparently spawned from the mailing list of the first one. If every new catalog sells its mailing list to five other companies, and it takes a month for the new company to deliver a catalog and sell the mailing list, how many catalogs total will Millie receive by the end of the year?

6.   THE ANTS COME MARCHING IN

I left out some pizza last night, and I'm paying for my mistake this morning. When I got up I counted 50 ants. I followed their trail back and saw that more were coming. In fact, I noticed a group of 10. Next came a group of 13. After that, a group of 16 marched in. Each group was bigger than the previous group by 3 ants. This pattern continued through 13 more groups of ants. (All in all, I saw 16 groups of ants march in.) How many ants arrived in the last group?

7.   PARTY TIME

A home supplies manufacturer sells its products through household parties. Out of each party held, a skilled salesperson can develop 3 new parties for the next week. If a salesperson starts with 2 parties held the first week, how long will it take before she has held more than 300 parties?

8.   YOUR OWN SEQUENCES

Generate your own sequences. Write eight terms for each sequence.

  a. Generate an arithmetic sequence based on adding or subtracting the same value each time.

  b. Generate a geometric sequence based on multiplying or dividing by the same number each time.

  c. Generate a Fibonacci sequence.

  d. Generate a sequence based on an increasing difference each time (for example, add 1, add 2, add 3, and so on).

  e. Generate a sequence in two steps (for example, multiply by 2, subtract 1, multiply by 2, subtract 1, and so on).

©2004 Key College Publishing, *Instructor Resources: Crossing the River with Dogs*, Johnson/Herr/Kysh

Problem Set B, Version 2

1. SOMETHING BUT THE TRUTH

The police were having a hard time with a case. They caught three people at the scene of a store robbery. According to all evidence, there had been only two perpetrators: one main burglar and an accomplice. The third person was just an innocent bystander. Under questioning, the main burglar would obviously tell a lie, the innocent person would tell the truth, and the accomplice might lie or might tell the truth. From their statements below, determine which person was which.

Louise: I was the accomplice.

Manny: Rick is the main burglar.

Rick: Louise is the accomplice.

2. YALE RECORD CLUB

Yale Record Club made me an offer I couldn't refuse. For their low prices I could buy millions of records, tapes, and compact discs (CDs) and save millions of dollars. Well, anyway, I joined. The first month I bought 3 CDs, and the cost was $24.84. The second month I ordered 5 for a total of $38.82. Did I mention that the cost covered the CDs and the shipping and handling? Anyway, for $31.83 I bought 4 CDs the next month. I paid $17.85 for 2 discs the next month. I looked through their catalog and picked out 38 more CDs that I wanted to buy sometime. If I buy all 38 at once, then I pay the shipping and handling fee only once. On the other hand, if I buy them in smaller groups, well . . . What I really need to know is, how much would it cost to buy all 38 CDs as one order?

3. TEN POSTS

Arranged in a straight line are ten posts with 10 meters between consecutive posts. Mr. Jones must nail a sign on each of the posts. He wants to do as much walking as possible while doing this. He will only walk along the straight line of the posts. He is going to start at his home, which is 10 meters from the first post on the same line as the posts. He is carrying all ten signs. What is the length of the longest possible walk that will allow him to nail a sign on each post and return home?

Problem Set B, Version 2 *(continued)*

4. **THE PET STORE CONTEST**

The local pet store held a contest in which they drew ten children's names as semifinalists. (They gave each one a pet kitten—parental permission required.) They then held another drawing from among those ten to determine who would win the top two prizes: a pair of homing pigeons or a cockatiel. The ages of the semifinalists were 4, 5, 6, 7, 9, 10, 11, 13, 15, and 16. When the two winners' names were drawn, the older winner was a teenager. What is the probability that the younger winner's age was prime?

5. **WORLD WIDE WIDGETS**

In the Broadway musical *How to Succeed in Business Without Really Trying*, a young man works his way up from window washer to vice president of a company without doing much work. This problem is based on that musical.

At the company picnic of the World Wide Widget Corporation (offices in Topeka, Kansas), five men have gotten together. Their names are Gus, Woody, Ned, Dick, and Jake. They all live in Topeka now but grew up in five other Kansas towns: Wichita, Dodge City, Belleville, Concordia, and Lawrence. Their jobs with the company are (in order from lowest to highest ranking) window washer, mailroom clerk, bookkeeper, manager, and vice president. Some of them were hired at the absolute lowest position of window washer and worked their way up, while others were hired at some higher level and worked their way up. In any case, each of them worked at each level for one year and then moved up to the next level. (As you might expect, this company had lots of vice presidents.) From the clues below, determine each man's name, position with the company, and where he grew up. (*Note:* It is possible to play more than one game at the picnic.)

1. The man from Belleville works in a position immediately higher than the man from Lawrence. They are each attending their first company picnic. Vice president Gus has been playing softball with them.

2. Gus and Ned played horseshoes with the man from Wichita at last year's picnic.

3. Jake has been playing basketball with the window washer, who is attending his first picnic.

4. The man from Dodge City will be a vice president next year. He and Woody ate lunch together at last year's picnic.

©2004 Key College Publishing, *Instructor Resources: Crossing the River with Dogs,* Johnson/Herr/Kysh

Problem Set B, Version 3

1. BURBANK NEIGHBORS

There are exactly five houses that occupy the entire length of one block. In front of each house is a car belonging to the owner of the house. The five owners are Stanley, Neuerburg, Stump, Frick, and Wahhab. The colors of the cars are blue, green, maroon, red, and yellow. Determine who lives in which house and which car each drives.

1. The person who owns the fifth house owns the yellow car.

2. The house Neuerburg owns has houses on both sides of it.

3. Stanley owns the maroon car.

4. The person who owns the second house owns the red car.

5. Wahhab owns the middle house.

6. Stump does not own the yellow car.

7. The person who owns the first house owns the green car.

2. BUCKS FOR CLUCKS

Monty loves game shows on TV—so much so that I promised I would go with him to a taping of *Bucks for Clucks* dressed as a farmer (Monty was dressed as a fox), and we would split whatever we earned down the middle. The costumes were apparently good enough, because we were chosen to "crack the golden egg." There were three huge fiberglass eggs on the stage, each painted gold and each with some sort of prize in it. (There is always one "turkey," a good prize, and a great prize. The real golden egg is the one that contains the great prize.) Each egg had a sign in front of it.

Egg 1: The turkey is in egg 2.

Egg 2: This egg is the golden egg.

Egg 3: The turkey is in egg 1.

As usual on this show, we knew that one of the signs was accurate and the other two were incorrect. (In the parlance of the show, the other two were "rotten.") Which egg should Monty and I choose in order to maximize our chances for getting the great prize?

Problem Set B, Version 3 *(continued)*

3. KANGA AND ROO

Kanga and Roo decided to have a race. Kanga jumps 8 feet with every jump and makes one jump every 4 seconds. Roo, on the other hand, jumps twice as fast (one jump every 2 seconds) but jumps only 4 feet with each jump. The racecourse was 100 feet long, with the race being up and back. (So that's 200 feet total.) Who won and by how far?

4. DOG AND TREES

We had just moved into our new home in the country. I took my favorite howling dog, Spam, out so that he could sniff around the yard. Well, dogs being territorial, he wasn't content just to sniff. He marked off his territory. There was a windbreak at the edge of our property with 12 trees in a straight line, all about 8 meters from one to the next. I'm sure that ol' Spam walked to and marked each and every tree. Furthermore, as he went from one tree to another, he walked the greatest distance possible. He didn't just go from one tree to the next, but he went to some other tree in the windbreak so that he never marked the same tree twice and he walked the longest total distance possible. How far did Spam walk from whatever tree he started at to whatever tree he ended at?

5. PRODUCT OF OUR TIMES

I think it was back in 1990 that I wrote the date 3/30/90 and noticed that the month times the day gave the year. I noticed it again on October 9: 10/9/90. That made me wonder: How many times did this occur during the decade of the 1990s?

©2004 Key College Publishing, *Instructor Resources: Crossing the River with Dogs,* Johnson/Herr/Kysh

Problem Set B, Version 4

1. NEW CAR OPTIONS

June wants to buy a new car. She visited her local new-car dealer and found that he has cars with exterior colors of (in order from darkest to lightest) black, navy blue, burgundy, green, tan, and white. For the interior color she can order any of those colors, but the interior color must be the same as, or lighter than, the exterior color. However, no cars combine green and burgundy or green and navy blue. Hood ornaments may be ordered for cars that are burgundy-colored (outside, or inside, or both). Sunroofs may be ordered for cars with white or tan interiors. Fancy hubcaps are available for cars with black or navy-blue exteriors. How many different styles of cars are available?

2. SHEET CAKE

Russell's cake measures 18 inches by 20 inches. The cake sat on the table in front of Russell with the 18-inch sides on the left and right. Russell cut off pieces of the cake, each measuring 2 inches by 2 inches, and served them to the guests at his party. His friends all like sugar, so he cut off side pieces, starting at the bottom right, and worked his way clockwise all the way around the cake. After he cut off the last side piece, he continued cutting and serving in a clockwise spiral starting from the bottom right. What was the original location of the last piece he served?

3. PLANET THREA

On the planet Threa, the days have 24 hours and there are 60 minutes per hour. The 12-hour clocks go from 1 o'clock to 2 o'clock, and so on up to 12 o'clock, and then start again at 1 o'clock. The digital clocks are different from Earth's, however. The minutes come before the hours. So the time that reads 4:10 is actually 4 minutes after 10, a time that on Earth would appear as 10:04. A visitor from Earth could get quite confused at certain times. Yet the Threa time 21:06 would not be confusing, since you would never see a time that looked that way on a 12-hour clock on Earth. What is the probability that an Earth visitor looking at a digital clock on Threa will see a time that could represent a valid time on a 12-hour digital clock on Earth?

4. CAMP SACRAMENTO

At Camp Sacramento, there was a great playground. Five kids loved to play there: Lisa, Danny, Justin, Jacob, and Jamie. Their last names were Bland, Walker, and Horlick (there were two sets of siblings among the five). Each child had a favorite activity at the playground: slide, tire swing, monkey bars, rings, and a big rock. From the following clues, determine each person's full name and favorite activity.

(This problem is continued on the next page.)

1. At Camp Sacramento, there were three groups for the kids: the Marmots for kids 10 to 12 years old, the Chipmunks for 6- to 9-year-olds, and the Minnows for 3- to 5-year-olds. None of the five children were Minnows. Jacob and the younger Horlick were Chipmunks. Lisa and the two Walkers were Marmots.

2. Lisa didn't spend very much time on the playground, but when she did, she never went on the slide or the tire swing.

3. None of the Marmots liked the rock. Danny said he didn't either.

4. Jamie was only 9, but she was a Marmot anyway so that she could be in the same group with her brother.

5. The elder Horlick and the two Chipmunks didn't like the monkey bars. A girl liked the tire swing.

5. **PLAYING DETECTIVE**

Four suspects were assembled in the director's office, having been accused of a devious crime: turning off the light switch during Mr. Buehler's business presentation. It was known that only one of the four turned off the switch. All four were friends, and the director's secretary overheard them plotting before they were brought into the director's office. They all agreed to tell the same number of false statements, although the secretary did not hear the agreed-upon number. Their statements are below. Who turned off the light switch?

Joe: Frank didn't do it.
I went to college with Felipe.
I didn't do it.

Felipe: I didn't do it.
Joe didn't go to college with me.
John didn't do it.

John: We all agreed to tell one false statement.
I didn't do it.
Felipe did it.

Frank: We all agreed to tell two false statements.
Felipe didn't do it.
I didn't do it.

6 Guess and Check

Guess-and-check is an effective, powerful problem-solving tool. It is a relatively newly articulated strategy and, as such, faces a lot of obstacles. Many people who are trained in mathematics have a difficult time accepting guess-and-check as a valid problem-solving strategy that students should know. Although it is appearing in new editions of many textbooks, it's often presented as a sidelight, not as an essential strategy for solving problems in the text. In time, guess-and-check will be more universally accepted, and the method for teaching the strategy will be framed and refined.

Although we were generally taught in our own math classes not to guess, this strategy involves more than guessing. It is a strategy of guessing and then guessing again. The guesser evaluates each guess according to the conditions of the problem and records the results in an organized fashion. Subsequent guesses are not so much guesses as they are guesstimates. The strategy might be described as "guesstimate, compute, organize, evaluate, and refine." Each successive guess develops more information about the problem or develops more data about the solution.

In order for students to learn from their guesses, the guesses must not be random. Students must organize the guesses and computations in a chart. There is, however, no set formula for setting up the chart. In fact, you need to impart to students that a chart that doesn't seem to be doing the job needs to be scrapped. In general, having too many columns in a chart is better than having too few, even if some of those columns contain information that remains constant through the whole problem.

Organizing the guesses and computations into a chart allows students to find more information in the problem and will help them better understand the problem. The inherent patterns in the problem will become more evident, and students can use these patterns to move toward the solution. Students may reach the answer through guess-and-check. Or they may learn enough about the problem to either start over or abandon guess-and-check in favor of a different strategy, such as algebra or working backwards. (Algebra as a problem-solving strategy, including converting a guess-and-check chart into algebraic equations, is covered in Chapter 13: Convert to Algebra. The strategy of working backwards is covered in Chapter 11: Work Backwards.)

You need to model refined guesses so that students learn to organize their guesses in a systematic way. Discuss what you learn about the problem as you go through it and explore how your guesses help you understand the problem better. Model writing out complete charts and "bracketing," surrounding the solution with a high and a low guess. Students tend to skip writing things down because of overconfidence, so be careful not to model overconfidence. They will also need to see how much more effective it is generally to start with small numbers instead of big numbers.

Many students will resist learning guess-and-check. This will be especially true of students with a good algebra background. Although we encourage divergent thinking as a central aspect of problem solving, this chapter is not the place for students to diverge. Some will use algebra as a way to avoid learning this new strategy.

Guess-and-check is also a valuable tool in the real world. Even people who have mastered algebra find guess-and-check useful; the way guess-and-check mimics algebra helps them remember algebra they may have forgotten.

Guess-and-check needs lots of modeling and practice. For this reason, Problem Set A contains a lot of problems. You can also use Problem Set A, Version 2, if your students need further practice.

One fun way to introduce guess-and-check is with the 2-4-6 Game, taught to us by Tom Sallee. The leader of the game thinks of a rule for three numbers that 2-4-6 always satisfies. Then the class asks if different series of three numbers satisfy the rule. The leader responds yes or no, and the class tries to determine from these responses what the rule is. Here's an example:

2-4-6	yes	7-6-4	yes
2-5-6	no	6-7-4	no
2-3-6	no	6-4-0	no
2-4-7	no	6-4-2	yes
3-6-6	yes	8-5-5	yes
5-4-6	yes	8-5-7	no
4-2-6	yes	8-4-7	yes
800-4-6	yes	8-1-9	yes

In this example, the first two numbers form a number that is a multiple of the third number. For example, 24 is a multiple of 6.

6 · GUESS AND CHECK

©2004 Key College Publishing, *Instructor Resources: Crossing the River with Dogs*, Johnson/Herr/Kysh

Good Quiz Problems

For a one- or two-problem quiz on this strategy, consider the following problems from Problem Set A, Version 2. For a two-problem quiz, we suggest one coin problem and one other problem. Consider using the following directions.

Solve each problem by guess-and-check. Show neatly organized charts with accurate labels. Each chart must include at least three guesses. Clearly state your answer. You will be graded 2 points for trying, 2 points for understanding, 4 points for the quality of your strategy, and 2 points for the answer clearly stated. You do not have to explain anything. But be sure your guess-and-check charts include all of the components we discussed in class. You need to show at least three guesses. At least one guess must be high, and at least one guess must be low.

1. Quarters, Dimes, and Nickels

2. Change

5. Health Club

6. T-Shirts

9. Compact Discs

11. Lots of Coins

12. Change in the Drawer

13. Broadway Boutique

14. Fitness Costs Money

Problems like 4, Long Journey, are good problems for the midterm.

Text Problems

SATURDAY AT THE FIVE-AND-DIME GARAGE SALE

Cinci held a garage sale, during which she charged a dime for everything but accepted a nickel if the buyer bargained well. At the end of the day she realized that she had sold all 12 items and had a total of 12 nickels and dimes. She had raked in a grand total of 95 cents. How many of each type of coin did Cinci have?

Do this problem before continuing. Even if you think you can write an equation for it, don't. Solve it by guessing a possible answer and checking to see if that answer is correct. Then make another guess, check it, and so on.

©2004 Key College Publishing, *Instructor Resources: Crossing the River with Dogs*, Johnson/Herr/Kysh

■ **FARMER JONES**

Farmer Jones raises ducks and cows. She tries not to clutter her mind with too many details, but she does think it's important to remember how many animals she has and how many feet those animals have. She thinks she remembers having 54 animals with 122 feet. How many of each type of animal does Farmer Jones have? Do this problem before continuing.

©2004 Key College Publishing, *Instructor Resources: Crossing the River with Dogs*, Johnson/Herr/Kysh

■ **ALL AROUND THE PLAYING FIELD**

The perimeter of a rectangular playing field measures 504 yards. Its length is 6 yards shorter than twice its width. What is its area? Solve this problem before continuing. You might find that a diagram is helpful as well as guess-and-check.

©2004 Key College Publishing, *Instructor Resources: Crossing the River with Dogs*, Johnson/Herr/Kysh

■ **CASCADES STATE PARK**

Emi and Margit had stopped at the bottom of one of the highest waterfalls in Cascades State Park. As Emi looked up at the waterfall, she said, "Wow, I think the top of that fall is about 20 feet more than three times the height of that young redwood!" Margit, of course, had a different opinion. She said, "No, I think it's about 50 feet less than four times the height of the redwood." If both are approximately right, about how tall is the redwood and how high is the waterfall? Work this problem before continuing.

©2004 Key College Publishing, *Instructor Resources: Crossing the River with Dogs*, Johnson/Herr/Kysh

Text Problems (continued)

■ **THE MONA AND LISA PAINTING PROBLEM**

Working alone, Mona can paint a room in 4 hours. Working alone, Lisa could paint the same room in 3 hours. About how long should it take them to paint the room if they work together? Show the answer to the nearest tenth of an hour. Work this problem before reading on.

■ **NEXT TRAIN EAST**

A train leaves Roseville heading east at 6:00 a.m. at 40 miles per hour. Another eastbound train leaves on a parallel track at 7:00 a.m. at 50 miles per hour. What time will it be when the two trains are the same distance from Roseville? Do not read on until you've worked this problem.

Problem Set A, Version 2

1. QUARTERS, DIMES, AND NICKELS

Jon has three times as many dimes as he does quarters. He has as many nickels as he has dimes and quarters combined. The total amount of money he has is $3.00. How many of each coin does he have?

2. CHANGE

Shadi has $4.60 in quarters, dimes, and nickels. She has three more dimes than nickels and three more quarters than dimes. How many of each coin does she have?

3. ROCKS

Christopher has four more than twice as many rocks in his wagon as Gordon has in his wagon. If he gives Gordon six rocks, he will have one more than Gordon. How many rocks does each boy have in his wagon right now?

4. LONG JOURNEY

Maureen drove 50 miles per hour to her sister's house. The two of them drove from there to their mom's house. Maureen's sister, Leann, drives a tad faster than Maureen (at about 60 miles per hour). Maureen took a total of 4½ hours to travel 255 miles from her house to her mom's house by way of Leann's house. How far does Maureen live from Leann?

   Alternate question: How long did it take Maureen to drive to Leann's house?

5. HEALTH CLUB

Estela's current health club is raising their usage fees to $38 per month. Their membership fee is $160, which is a one-time-only charge. Estela is considering joining another club, but she'd have to pay a membership fee of $250 and then usage fees of $32 per month. How many months will it take before the new health club is the cheaper plan in terms of total cost?

6. T-SHIRTS

Blaise had a clerk use a markup of 25% on one line of T-shirts at his store. The selling price is $9.75. What was the price of the shirt prior to being marked up?

Problem Set A, Version 2 *(continued)*

7. CENTRAL VIRGINIA COLLEGE

Central Virginia College has about 6 sophomores for every 7 freshmen. There are currently 1131 students in those two classes. How many more freshmen are there than sophomores?

8. MUTUAL FUNDS

As an investment principle, Hampton National Mutual Funds tries to keep a ratio of $9 blue-chip stocks to every $2 in high-risk stocks. If the company currently has $95,700 invested in those categories of stocks, how much does it have invested in blue-chip stocks?

9. COMPACT DISCS

Keith wants to join Liberty Record Club, which is offering new members a $20 membership fee and a cost of only $6.20 per compact disc. On the other hand, Patriot Music Club charges no membership fee and the cost of each compact disc is $8.10. Deanna wants to join Patriot. Keith and Deanna are married, so they don't need to join both clubs. How many compact discs would they need to buy before Liberty would be cheaper?

10. STICKERS

Cici and Amatina have a lot of stickers. Cici had one-third as many as Amatina had, but then Amatina gave her six stickers, so now Cici has half as many as Amatina. How many stickers did each girl start with?

11. LOTS OF COINS

Caleb has $10.95 in quarters, dimes, and nickels. He has four more dimes than quarters. He has a total of 80 coins. How many of each coin does he have?

12. CHANGE IN THE DRAWER

Karin has $8.15 in quarters, dimes, and nickels. She has three more nickels than quarters. She has a total of 70 coins. How many of each coin does she have?

TEACHING RESOURCES

©2004 Key College Publishing, *Instructor Resources: Crossing the River with Dogs,* Johnson/Herr/Kysh

Problem Set A, Version 2 *(continued)*

13. BROADWAY BOUTIQUE

Candice and Marina own The Broadway Boutique, a store that sells various items from Broadway shows. All of the items in their store are marked up 40%. In other words, they take the amount they pay for an item and mark that price up 40% to get the price they charge their customers. Their newest items are the little green plants featured in *Little Shop of Horrors.* They sell each plant to their customers for $26.18. What was their cost for each plant?

14. FITNESS COSTS MONEY

Courtney wants to join a health club because working out will help her golf game. She can join Big Muscles Health Club, which has a one-time membership fee of $140 and charges $46 per month. Her other choice is Being Fit Health Club, which has a one-time membership fee of $270 and charges $37 per month. How many months will it take before the total cost of Being Fit is less than the total cost of Big Muscles?

©2004 Key College Publishing, *Instructor Resources: Crossing the River with Dogs,* Johnson/Herr/Kysh

Problem Set B, Version 2

1. SPARE CHANGE

Back in 1994, a cashier found that he was often asked to give change for a dollar to people who had made no purchase but wanted 20 cents for a telephone call. He started thinking one day about the number of ways he could make change. If he gave no more than five of any type of coin, used no 50-cent pieces, and made sure that the person received coins to make exactly 20 cents for the phone call, in how many different ways could he give change for a dollar?

2. SKI TRIP

A group of friends decided to rent a house in Aspen, Colorado, for a week of skiing. They each had to chip in $70 for the week's lodging. If they had been able to convince three more people to go, the cost per person would have been reduced by $14. What was the rent for the week?

3. EXPENSIVE MISSILE

Some senators were sitting around discussing the latest Defense Department budget. Several of them asked the vice president how much the latest missile system cost the taxpayers. The vice president, who was known to be quite clever, wouldn't tell them straight out but instead gave them clues. (Note that the first digit is on the left.)

The number is ten digits long (there are no decimal points), and each digit is different.

The first, third, and fifth digits are powers of 3.

The first, second, third, fifth, and ninth digits are odd.

The first, second, seventh, and ninth digits are prime.

The sum of the seventh and ninth digits is the third digit.

The first digit is not 1, the second digit is not 2, and so on with no digit occupying its numbered place, up to the tenth digit, which is not 0.

No number is adjacent to a consecutive number. So, 1 is not next to 0 or 2, 2 is not next to 1 or 3, and so on.

How much did the missile system cost?

TEACHING RESOURCES

©2004 Key College Publishing, *Instructor Resources: Crossing the River with Dogs*, Johnson/Herr/Kysh

Problem Set B, Version 2 (continued)

4. FROG CHORUS

Behind my house is a big field. Every winter, after a few rainstorms, the field seems to attract lots of puddles, mud, and frogs. The frogs are quiet for the most part. However, when one starts croaking, they all do, which serves to attract more frogs. And then they all seem to shut up at once. One night, 30 frogs were present, and they all started croaking at 11:00 p.m. They all croak for a number of seconds equal to the number of frogs present. After that amount of time, 1 frog drops out at the beginning of every subsequent second. They all stop abruptly after the second in which they realize that only half of them are croaking. Two more frogs arrive every 20 minutes, and this starts the whole process over again with massive croaking. (Every 20 minutes means that new frogs arrive at 11:20, 11:40, 12:00, 12:20, and so on.) How many seconds of croaking took place from when I went to bed at 11:00 p.m. (and they started croaking) until I got up at 6:50 a.m.?

5. THE AYER FAMILY

Lawrence and the three other Ayer children each have birthdays in a different month. One birthday is in August, another is in January. This year all of the birthdays fell on a different day of the week (one was on a Saturday). The ages of the children at the beginning of the year were 7, 11, 14, and 15. Determine the month and the day of the week (in that year) of each child's birthday and how old each was at the beginning of this year. Note that a week is considered to begin on Sunday and end on Saturday.

1. Last year (a leap year) two of the birthdays were on Tuesdays, but this year none of the birthdays were on Tuesday, although one birthday was on a Wednesday.

2. This year the November birthday was not on a weekday.

3. Don and the person with the Monday birthday were the same age for more than two months during the year.

4. The person with the Thursday birthday was born four years earlier than Paul.

5. The July birthday came later in the week than Lucy's.

Hint: Consulting a calendar may help you solve this problem.

Problem Set B, Version 3

1. READY FOR "THE SHOW"

Carl is vain. He is so vain that he would probably think this problem is about him. It's not, but let him think that anyway. Carl wants to be a major-league baseball player. Every day before he goes out in front to practice hitting on a tee, he puts on one of his baseball outfits. He chooses a hat, socks, and shirt that all go well together. He has a blue hat, a red hat, and a black hat. He also has a green shirt, a yellow shirt, and a red shirt. And Carl has baseball socks in each of the same colors that his shirts and hats come in, except black. He also has a pair of orange socks. He never wears more than two colors at a time. He also believes that blue and red together are "bush league" (they don't go together well) and that red and green belong together only in the "off-season." He also refuses to wear yellow and black at the same time, because that combination makes him feel like a bumblebee. What is the probability that he wears something green?

2. BIKE RACE

The organizers of a long-distance bike race wanted to set up food and aid stations every 8 miles along the route. There would be no aid stations at the start and the finish. Each station would be staffed with one volunteer to give out food and dispense aid. Unfortunately, on the day of the race, three of the volunteers didn't show up. The organizers quickly determined that everything would work out if they placed the aid stations 10 miles apart instead of 8 miles apart. How many volunteers showed up?

3. STORKE TOWER

Storke Tower, the bell tower of the University of California at Santa Barbara, chimes every hour at 10 minutes before the hour (to remind students that they should be getting to class) and also on the hour (to tell them that they are late). The tower chimes 24 hours a day, but it uses a 12-hour clock. (So, for example, it chimes 12 times at midnight and at noon.) Every time it chimes, it plays a 16-note tune, which is broken into four parts of 4 notes each. Each note lasts 1 second, and there is a 1-second pause at the end of each 4 notes. In addition to this, when it chimes on the hour, it also chimes out the time. So at 10 o'clock it would chime 10 times more after going through its 16-note tune. When it chimes for the hour, the chimes last 1 second, and there is a 1-second pause between chimes. There is a 3-second pause between the end of the 16-note tune and the beginning of the hour chimes. How much time every day does the bell tower spend chiming, including the pauses between chimes?

TEACHING RESOURCES

©2004 Key College Publishing, *Instructor Resources: Crossing the River with Dogs,* Johnson/Herr/Kysh

Problem Set B, Version 3 (continued)

4. COLLEGE ROOMMATES

Helen and her three roommates live in a two-bedroom apartment at The University of Texas at El Paso. Each has a different major (one woman is majoring in chemistry). Coincidentally, each woman also has a different hair color (one is blonde). Determine each woman's full name, hair color, and major.

1. Joan, who isn't Ms. Bonds, doesn't have either of the two darker hair colors (black or brown), and neither does Irene.

2. Glory doesn't share a room with Ms. Carlson or the accounting major.

3. Glory's major isn't physics.

4. The physics major has brown hair.

5. Ms. Daniels's hair is neither red nor black.

6. Ms. Bonds, whose first name is not Glory, is majoring in English.

7. Ms. Alder and Helen are both majoring in a science and often take classes together.

5. LUNCH MONEY

Cornelius and Sally are husband and wife. One day, they were discussing finances. Sally accused Cornelius of spending too much money on lunch.

Cornelius said, "But I really haven't spent that much money on lunch recently."

Sally replied, "Suppose you tell me how much you spent on lunch each of the last three days."

"Okay, but I will make you figure it out. Each day I spent a whole number of dollars. If you multiply the amounts I spent each day, the answer is 96."

Sally thought for a while and then said, "I need more information."

Cornelius answered, "The total amount I spent for the three days is the same amount as what you spent on golf last weekend."

Sally knew how much she had spent for golf, but that still wasn't enough information for her to figure out the answer. She asked for another clue.

"Well, two of the meals cost less than $5."

That proved to be enough for Sally to figure out how much Cornelius spent on each meal. You figure it out too.

Problem Set B, Version 4

1. GRANDPA'S ATTIC

We always liked poking around Grandpa's attic whenever we had a family reunion. We found all sorts of neat stuff up there. Once we found a bunch of baseball cards, so Grandpa said, "Just divide 'em up among all the grandchildren." There were 5040 cards in all, so each of us got a lot of cards. But then we remembered that the Yakliches, who had five of the grandchildren, hadn't arrived yet. So each of those of us present had to give up 75 cards so that all the grandchildren would have the same number of cards. How many grandchildren does Grandpa have?

2. THE MATH TEST

Violet was taking a math test. On one problem she was given two positive whole numbers. The problem asked her to square the two numbers and add the squares together. Unfortunately, Violet misinterpreted the question. She mistakenly added the two numbers first and squared the result. Her answer was 60 greater than the correct answer to the question. Find all possibilities for the original two numbers.

3. FAMILY OUTINGS

The Parker family plans four major outings every year. The outings take place on major holidays: Presidents' Day in February, Memorial Day in May, Independence Day in July, and Labor Day in September. Each outing (one trip was to the county fair) is planned by one of the four Parker kids (one's name is Andy). During the outing the Parkers eat lunch (once they ate deli sandwiches). Determine which child planned which holiday's outing, where they went, and what kind of food they ate.

1. The trips, in chronological order, were Erin's outing, the zoo trip, the time they ate hamburgers, and Neal's outing.

2. They ate pizza after Presidents' Day but before Jenny planned her outing.

3. The Parkers went to the museum on the Fourth of July (Independence Day).

4. They ate hot dogs at the amusement park but not on Labor Day.

4. RHO-SHAM-BO

Griffin, Craig, and Tony are playing rho-sham-bo, otherwise known as rock-paper-scissors. It is a game for two players. The game has gotten boring because Griffin always chooses rock, Craig always chooses scissors, and Tony always chooses paper. Thus, when Griffin plays against Craig, Griffin wins (rock smashes scissors). When Craig plays Tony, Craig wins (scissors cut paper). And Tony beats Griffin when they play (paper covers rock). Suddenly, two of the participants exchange preferences, while the third participant remains the same. If Craig now plays Griffin, who wins?

5. POOR SCORE

Bob and Bob played golf against each other in a tournament. A marshall keeping their score had a difficult time because both players were named Bob. The scores the marshall recorded were the correct scores, but they may have been reversed. This is the way the marshall recorded their scores:

Hole	1	2	3	4	5	6	7	8	9	Total
Par	4	4	5	3	5	4	3	4	4	36
Bob A	3	4	4	2	5	3	4	4	3	32
Bob B	4	4	7	4	3	4	3	5	6	40

When the match was over, the two Bobs glanced at the scorecard and complained.

Bob B said, "Wait a second, I had only one double bogey. And there is no way I lost by eight shots: I had fewer total shots until after we played the fifth hole."

Bob A said, "I had the eagle, but I had only two birdies. I won only three holes."

Determine the correct hole-by-hole score for each player and their totals for the nine holes.

(*Note:* Eagle, birdie, bogey, and double bogey refer to scores made on one hole. An eagle is two under par for the hole. A birdie is one under par. A bogey is one over par. A double bogey is two over par. A player wins a hole from the other player when he has the lower score for that hole.)

7 Identify Subproblems

Solving subproblems is a strategy based on dividing problems into smaller pieces. A subproblem is a mini-problem that you must solve before solving the original problem. Some problems break down into many subproblems. To solve the type of problem that breaks down into subproblems, it is often helpful to list the subproblems. This list then becomes a plan for solving the problem.

Solving subproblems is a lot like building a bicycle. It doesn't matter much whether you attach the front tire or the rear tire first. It also doesn't matter much if you attach the seat or handlebars before, after, or while you're attaching the tires. However, for the bicycle to be complete, each of the smaller steps must be completed. The order of many of the steps is not necessarily important. But for other steps, the order may be very important. For example, you must attach the rear tire and the pedal crank before you put on the chain. The same is true when you solve a problem with subproblems—you must solve some subproblems before others; in other cases, the order is irrelevant.

You or your students may also encounter situations where you can't solve a problem immediately, because you need more information. Breaking problems down into subproblems makes this more readily apparent.

In the larger scheme of the major problem-solving themes, solving subproblems is a strategy that requires you to change your focus. No longer are you looking at a large problem. Rather, you must delve into the interior of the problem to find the component parts—the subproblems.

Solving subproblems also has aspects of an organizational strategy, although you organize your *approach* rather than your information.

Solving subproblems is a useful strategy for real-life problem solving. Beyond tests that have a strong emphasis on subproblems (such as the SAT), much of human activity is complex enough that it can be viewed as subproblems. Manufacturing a bicycle is an example; beyond the manufacturing aspect, there is also purchasing, training, and marketing to consider. The manufacturing stage is simply a subproblem of the larger problem of making, distributing, and selling bicycles.

Notes on Text Problems

PAINT

This is a mixture problem of the type you may see in an algebra text. Notice, however, that algebra is not needed to solve it.

CHOCOLATE MILK

This is another typical algebra mixture problem. The text recommends using a combination of two strategies to solve it: subproblems and guess-and-check.

Good Quiz Problems

For a one-problem quiz on this strategy, consider one of these problems from Problem Set A, Version 2:

1. Magnet School

14. Classic Books

15. Long Road

16. Cat Loves Beanie Babies

17. Honor Society

©2004 Key College Publishing, *Instructor Resources: Crossing the River with Dogs*, Johnson/Herr/Kysh

Text Problems

AT THE STUDENT BOOKSTORE

Shanein is a student at Oregon State. She has four $20 bills. She buys six notebooks at $3.95 each, but they are marked 20% off. She also buys three highlighters at $1.19 each and a textbook at $44.98. There's no sales tax in Oregon. How much change should Shanein receive? Work this problem before continuing.

WATERING THE LAWN

Three quarts of water are needed to water 1 square foot of lawn. How many gallons of water are needed to water a lawn that measures 30 feet by 60 feet? List the subproblems and answer the question before proceeding.

THE CAR BARGAIN

Paul went into the local new-car lot to buy a car. He knew the kind of car he wanted, because his friend Barbara Gain had bought the same car the day before. Barbara received a 30% discount on the car, which listed at $15,000. The salesperson offered Paul the $15,000 car at a 20% discount instead. When Paul protested, the salesperson offered an additional 10% off the 20% discounted price. This offer satisfied Paul and he bought the car, convinced he had paid the same price as Barbara. Had he? Solve this problem before continuing.

PAINT

A mixture is 25% red paint, 30% yellow paint, and 45% water. If 4 quarts of red paint are added to 20 quarts of the mixture, what is the percentage of red paint in the new mixture? List the subproblems and solve this problem before reading on.

Text Problems *(continued)*

CHOCOLATE MILK

Augustus is trying to make chocolate milk. He has made a 10% chocolate milk solution (this means that the solution is 10% chocolate and 90% milk). He has also made a 25% chocolate milk solution. Unfortunately, the 10% solution is too weak and the 25% solution is way too chocolaty. He has a whole lot of the 10% solution but only 30 gallons of the 25% solution. How many gallons of the 10% solution should he add to the 25% solution to make a mixture that is 15% chocolate? (Augustus is sure the 15% solution will be absolutely perfect.) Solve this problem before continuing.

©2004 Key College Publishing, *Instructor Resources: Crossing the River with Dogs*, Johnson/Herr/Kysh

©2004 Key College Publishing, *Instructor Resources: Crossing the River with Dogs*, Johnson/Herr/Kysh

Problem Set A, Version 2

Solve each problem by first listing all of the subproblems and then solving the subproblems to answer the question.

1. MAGNET SCHOOL

A magnet school program was set up to bring in students from all over the city. With the first 112 students, the magnet program had 50% from the south side of town, 12.5% from the north side, and 37.5% from the east side. If 6 more students from the north side and 2 students from the west side enrolled in the program, what would be the percentage of students from the north side of the city in the program now?

2. ORANGES

Oranges cost 50¢ per pound. How many oranges, each weighing 4 ounces, can Wally buy with five quarters?

3. BOYSENBERRIES

How much do 2 pounds of boysenberries cost if a 3-ounce basket of berries costs 54¢?

4. STEREO SALE

Williams & Son Department Store was going out of business. At first they had all of the electronics marked at 10% off. They then marked everything down an additional 25% from the last posted price. What is the current price of a stereo set that originally cost $189?

5. LIFE'S NECESSITIES

Jack and Tim have decided to run away, but they need proper sustenance. They have enough money to buy 30 donuts or 20 sodas. Jack told Tim that he thought they needed only 18 donuts. How many sodas can they then buy?

6. AN A IN MATH

Daniella wants an A in math. So far, she has 598 out of 700 points possible. What percentage of the remaining 225 points must she earn to raise her average to exactly 90%?

7. SOIL AND SAND

Eight cubic yards of soil that is 50% sand had to be mixed with some soil that is about 10% sand to lower the percentage of sand to 18%. From this action, how many cubic yards of soil with 18% were produced?

TEACHING RESOURCES

©2004 Key College Publishing, *Instructor Resources: Crossing the River with Dogs,* Johnson/Herr/Kysh

Problem Set A, Version 2 *(continued)*

8. ONE HUNDRED SIXTY-EIGHT INCHES OF STRING

One hundred sixty-eight inches of string were used to make these squares, which are all the same size. What is the total area of the ten squares?

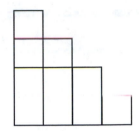

9. JUAN'S CHURCH

Juan donates 10% of his salary to his church. He earns $3,500 per month. He used to earn $3,200, and he gave the same dollar amount that he does now. What percentage of his income did he donate to the church when he earned $3,200?

10. A LOT OF STUFF

A patch of grass measuring 6 feet by 6 feet requires 2 ounces of fertilizer. How many pounds should be used on a lawn that measures 30 feet by 75 feet?

11. SHOOTING PERCENTAGE

Renee made 96 of 220 shots during the season up until tonight's basketball game. She figures she'll get about 20 shots tonight. What percentage must she shoot tonight to raise her shooting percentage to exactly 45% for the season if she takes 20 shots? (Shooting percentage is determined by dividing the number of baskets by the number of shots taken.)

12. CAMP STOVE

A camp stove uses up its gas in 195 minutes when it is set on low. It uses about 7½ times as much gas when it is set on high. The stove has been set on low for 30 minutes. How much longer can the stove operate if it must be run on high to boil water?

13. ROAD RALLY

In a timed road rally, Sandy and Debbie know that the winner is the person who comes up with the correct time over the entire course. The correct time, of course, is given by what time it would take to drive the course at the posted speed limits. The women have driven 5.5 miles in a 55 mph zone (freeway), 5 miles in 25 mph zones, and 3.5 miles in a 35 mph zone. They now see that the rest of the course is all in a 40 mph zone, all on a long country road. So far it has taken them 22 minutes. There are 6 miles remaining. How fast do they need to drive to come up with the correct course time? (Don't worry about the stop signs and stoplights—there aren't any.)

©2004 Key College Publishing, *Instructor Resources: Crossing the River with Dogs*, Johnson/Herr/Kysh

Problem Set A, Version 2 (continued)

14. **CLASSIC BOOKS**

Lucille has a large collection of classic books on four shelves in her office. The first shelf makes up 25% of the total and holds books written in the nineteenth century by Dickens, Hugo, and Bronte. The second shelf makes up one-eighth of the total number of books and holds Shakespeare's plays. The third shelf makes up two-fifths of the total and holds twentieth-century books by Steinbeck, Fitzgerald, and Hemingway. The fourth shelf holds 27 books, including mysteries by Hammett, Chandler, Christie, and Conan Doyle and science fiction by Asimov, Clarke, and Heinlein. How many books are on the third shelf?

15. **LONG ROAD**

Melody Road is a very long road in southern Slobovnia. It goes through four counties. Two-fifths of the road is in Staff County. Three-eighths of the road is in Note County. Ten percent of the road is in Flat County. Six miles of the road is in Tenor County. How long is Melody Road?

16. **CAT LOVES BEANIE BABIES**

Cleo the cat loves Beanie Babies. In the middle of the night she dragged all of the Beanie Babies from the three girls' rooms into the family room. All of the Beanie Babies were originally in the four bedrooms: Leslie's room, Allison's room, Laurie's room, and Gene's room. One-fourth of the Beanie Babies were originally in Leslie's room. Thirty percent of the Beanie Babies were originally in Allison's room. Three-eighths of the Beanie Babies were originally in Laurie's room. Cleo dragged all of those Beanie Babies into the family room. Cleo did not touch the 27 Beanie Babies in Gene's room. How many Beanie Babies were found in the family room in the morning?

17. **HONOR SOCIETY**

A regional honor society selected high school students from four cities. Fifteen percent of the students came from Rocklin. Three-eighths of the students came from Loomis. One-fourth of the students came from Roseville. There were 63 students from Auburn. How many students were from Roseville?

TEACHING RESOURCES

©2004 Key College Publishing, *Instructor Resources: Crossing the River with Dogs*, Johnson/Herr/Kysh

Problem Set B, Version 2

1. USED CARS

While watching a commercial on late-night TV, Bud noticed subtitles on the screen at the same time the announcer was saying something. "Sochi na warak" was on at the same time that the man said, "Sale cars cheap." Bud was intrigued, so the next several times he saw the commercial, he wrote down these subtitles and spoken phrases:

SUBTITLES	PHRASES
quandi bonki warak	big blue cars
quandi sochi	big sale
kandi sochi konaa	no sales people
warak na	cheap cars
bonki tayku ta	green and blue
konaa qualliak	no lemons
sochi na warak	sale cars cheap
wassa tayku	green money

Whatever language the subtitles are in, what is the meaning of the following sentence?

Qualliak na ta konaa tayku kandi.

2. THE NATURAL LOOK

A landscape architect worked on plans for the backyard of a client. She decided to use 15 trees and plant them in groups that were all different sizes to make the yard look "natural." The smallest group could have 1 tree in it, and there were at least two groups. With all possibilities considered equally probable, what is the probability that she included a group of 2 trees in her plans?

3. TYPESETTING

In the old days, type was set by hand with lead pieces. Each individual letter or number in each word had to be carefully put together. A printer wanted to print a list of whole numbers from 1 to 1000. If he had only 100 of each of the digits from 0 to 9, how many numbers could he set before he ran out of some digit?

©2004 Key College Publishing, *Instructor Resources: Crossing the River with Dogs,* Johnson/Herr/Kysh

4. **A FAMILY MAN**

Twelve years ago, a man was three times as old as his daughter. Nine years from now, the man will be twice as old as his son. Four years from now, the sum of the ages of the son and the daughter will equal the age of the man. How old is each now?

5. **HOW TO AVOID FALLING ASLEEP AT A MEETING**

I was sitting in a group of people arranged in a circle during a meeting. The meeting was rather boring, and I really wasn't paying attention to what was being said, so I began to count the people in the circle. I started with myself and went around the circle, counting by 4's, then by 5's, and so on, and looked at how many people were left over each time. The information is summarized in the chart below.

COUNTED BY	LEFT OVER
4	1
5	0
6	1
7	3
8	7
9	6
10	4
11	3

Somewhere in the middle of counting, I realized that 2 people were missing. I didn't know when they had left or if they had left together. When I finished counting by 11's, I noticed that they were both back. But I didn't know if they had come back together or if possibly one or both of them had come back and then had left again. In any case, I still thought I had enough information to figure out how many people were in the circle at the beginning. Did I? Can you? To help you out, there was a sign on the wall, posted by the fire marshall, limiting the capacity of the room to 80 people. How many people were in the circle, when did each of the 2 people leave, and when did they each come back?

Problem Set B, Version 3

1. WHO'S ON THE BENCH?

During a really boring baseball game, I started counting the guys sitting on the home-team bench. When I counted by fives, there were two left over. Counting by twos, there was one left over; counting by threes, there were two left over; counting by fours, there were none left over. That last bit sort of surprised me. I knew an odd number of players had been sitting on the bench because of the remainder from counting by two. But there was no remainder when I counted by four, which couldn't be correct. I sat there for a few minutes wondering what was going on. It suddenly dawned on me that the team had come in off the field. I looked up and saw two members of the home team out coaching the bases. There were two more guys (the batter and the guy on deck) who had never made it into the dugout. Nine guys had been on the field, but I didn't know at what point in my counting they'd come in. I counted the people sitting on the bench by sevens, and there was one left over. I knew there were fewer than 40 people on the team, including coaches. How many people had been sitting on the bench when I started counting? At what point had the team come in from the field?

2. GAMUSE PLAYGROUND

A Gamuse playground is a sight to see and an engineering wonder. This type of playground has 16 floors. Each floor is connected to another floor by various mechanical devices. Any even-numbered floor has an escalator up to the next even-numbered floor. Every odd-numbered floor has a slide down to the next odd-numbered floor. Several other floors (1, 2, 3, 5, 8, and 13) have an up-and-down "tube-chute" (like mail tubes in an office building) that will take you to another tube-chute floor above or below. How would somebody go from the eleventh floor of this playground to the twelfth floor?

3. BICYCLE TRAINING

Sherwood is a bicyclist. He makes a daily ride of 40 miles at a constant speed. If he trains hard, he can increase his speed by 4 miles per hour, which would result in a half-hour decrease in time. Find his new speed.

©2004 Key College Publishing, *Instructor Resources: Crossing the River with Dogs*, Johnson/Herr/Kysh

4. **SIERRA SLUGGERS**

Five members of the Sierra Sluggers went out to get something to eat after the last game of the season. Each one discussed the others' hitting:

Larry: I had 18 more hits than Jack.

Mark: The total for the five of us was 708 hits.

K. C.: Larry's total was a prime number, and he was second of the five.

Jack: Chris was between Mark and K. C. in the number of hits.

Chris: Mark's total was a multiple of 6, and K. C.'s was a multiple of 15. In fact, K. C. had 12 more hits than Mark.

Each person's statement was accurate. Determine how many hits each player had during the season.

5. **PRIZE MONEY**

Suppose the chamber of commerce in Pinehurst, North Carolina, decided to hold a professional golf tournament at the famous Pinehurst Country Club. They would probably be able to get a TV contract from a major network. The TV revenue would be about $4 million, and they would sell 60,000 tickets at $8 apiece. The fund for prizes would be 20% of their total revenue (TV plus tickets) rounded to the nearest $10,000. The prizes would be as follows (all prizes are rounded to the nearest dollar):

First prize: 18% of the total prize money

Second prize: 60% of the first prize

Third prize: 65% of the second prize

Fourth prize: 70% of the third prize

Fifth prize: 75% of the fourth prize

Sixth prize: 6% of the remaining prize money

Seventh prize: $900 less than the sixth prize

Eighth prize: $900 less than the seventh prize

And so on, with each subsequent prize $900 less than the previous prize. The final prize would be whatever was left over, even if it wasn't $900 less than the previous prize.

Suppose two golfers finished in a tie for tenth place, and they split the combined tenth and eleventh prizes. How much did each of them get?

Problem Set B, Version 4

1. MILEAGE SIGN

Recently I was driving down the freeway and spotted the following freeway sign with the distances to three upcoming cities:

Ahmanson: 147 miles

Chandler: 265 miles

Schubert: 380 miles

I thought the sign was unusual because the distances to the three cities featured all different digits. I realized that in 100 miles, the distances to the three cities would again have all different digits. But I wanted to know if that could happen sooner. In how many miles would I next see another freeway sign with all different digits for the distances to these three cities?

2. A NEW VERSION OF SCRABBLE

My friends have changed the rules for Scrabble. In their rules, each letter has a positive whole-number value. No two vowels have the same value. No two consonants have the same value. It is possible for a vowel and a consonant to have the same value. Scoring is done differently than in regular Scrabble. The points for the vowels are added, then (separately) the points for the consonants are added. These two numbers are multiplied together to give the word score. A list of words and their word scores is shown below. Determine the letter value of each letter, then determine the value of the word *problem-solving* (count this as one word and don't count the hyphen for anything).

PLUM = 36	PLUME = 48	MELT = 15
ROME = 98	SAVE = 51	WEST = 20
LAME = 18	PLANE = 39	MELBA = 51
PIPER = 110	VIPER = 150	WESTERN = 70
OUT = 81	GREET = 40	PAPER = 66

Problem Set B, Version 4 *(continued)*

3. **WIDE SLIDE**

At the local playground, three boys like to go down a very wide slide. After going down the slide, they turn around and climb back up the slide rather than walking around to climb the ladder. The slide is so wide that they can all slide and climb at once without crashing into each other. Each boy slides down the slide in two seconds. It takes Kasey five seconds to climb back up the slide. It takes Zac six seconds to climb back up the slide. It takes Denny seven seconds to climb back up the slide. They all start together at the same time and slide down. How many seconds after starting will they all slide down together at the same time again?

4. **THREE-DIMENSIONAL CHESS**

The three-dimensional chessboard is four squares by four squares on each level and is four levels high. In a particular version of three-dimensional chess, the knight moves by going straight up or down two levels, then forwards, backwards, right, or left one square on the new level. It can also move straight up or down one level, and then two squares (in a straight line) forwards, backwards, left, or right on the new level. It cannot stay on the same level for the entire move and act like the knight in regular two-dimensional chess. In the fewest number of moves, how can a knight move from a corner square on the bottom level to get to the corner square directly above it on the next level?

5. **LEAKY SINK**

My old house had a leaky sink. I used to shave in this sink by filling up the sink with water and then shaving while the water slowly leaked out. Sometimes the water would completely run out before I was finished. I made some calculations. It took 30 seconds to fill up the sink completely after I'd I turned on the tap. It took 3 minutes for the completely-filled sink to leak out and be empty. It took me 2 minutes to shave. I figured that the sink had to be at least 5% full in order to have enough water for me to shave effectively. What is the minimum amount of time I should turn on the water to be sure I have enough water to finish shaving without having to add more? (*Note:* I turn off the water before I start to shave, and I start shaving immediately upon turning off the water.)

TEACHING RESOURCES

©2004 Key College Publishing, *Instructor Resources: Crossing the River with Dogs*, Johnson/Herr/Kysh

8 Analyze the Units

Unit analysis is a problem-solving strategy you use to deal carefully with the units of a problem. The basis of this strategy is simply multiplying by 1. The key is to choose the correct form (or forms) of 1. The forms include fractions for which the numerators and denominators are equivalent, though they appear different. The purpose is to cancel the unwanted units and leave units that answer the question.

Units are a necessary component of a problem and its answer. When you solve any problem involving units, you need to be careful to maintain the units throughout the problem until they cancel. An answer cannot be correct without the correct units.

Keeping the units in a problem organized will help lead to the correct answer. In the extreme, it is possible to work a problem, with no sense of what the problem is about, simply by using the units to move from problem to answer. We don't mean to endorse such a shallow manipulation, but applying this strategy can serve as a means for checking one's work. The units should be an integral part of the means for solving the problem and also an indicator of what ends (correct units) to work toward.

Unit analysis is a strategy that organizes information. Problems seem far less formidable when information is organized. With organized units, problems should simply fall into place. However, teach students to also analyze what types of units are presented in a problem. In other words, what is being measured: length, mass, time, volume?

This chapter is divided into three sections: Units in Ratios, Unit Conversion, and Compound Units.

In the Units in Ratios section, the examples are kept to simple, familiar ratios to develop the concept of unit ratios. For example, miles per hour is usually an easy concept for students to dissect, and we have found that it is an extremely good example to start with. From there, we proceed to more abstract concepts, though usually with some intermediate discussions. Another good example for students is 30 jellybeans for $1.50, because you start with jellybeans per dollar and you can change it into cents per jellybean and then into dollars per jellybean. The unit changes are easily done, and the results make sense to the students because of their prior experience. Miles per gallon and dollars per gallon also work well, and unusual ratios can be created easily, such as socks per person or bones per dog.

In the Unit Conversion section, a unit conversion simply involves changing units of measure within the English and metric systems to other units of measure within the same systems. In the English system we can convert from feet to yards, or feet to inches, or even feet to miles. Conversions in the metric system are simpler: from meters to kilometers or perhaps from meters to centimeters. Converting units also involves converting from metric to English or vice versa.

When doing metric-to-English conversions, you do not need to develop a lot of different conversion ratios. Using just two conversion ratios is sufficient for the students to readily go back and forth between the measurement systems:

Linear measure: 1 meter = 3.281 feet

Volume: 1 gallon = 3.79 liters

The appendix at the back of the student book contains more handy information like this about measurement, as well as a review of operations with fractions.

In the Compound Units section, familiarity is again a key. Though students are generally familiar with the concept of square feet, their depth of understanding is often severely lacking. The idea of passenger-miles is also within the grasp of most students because both units are so familiar. Compound units are extremely important in other disciplines, especially in physics.

Students also need to be aware that solving subproblems is a major component of the strategy of unit analysis. Subproblems show up quite naturally in a good number of the problems in this chapter. In many problems you will be unable to go directly to the desired units, so you must set up intermediate goals.

Though multiplying by 1 is fundamental to doing operations with fractions (since 1 is the identity element for multiplication), many students will not have a deep understanding of the unit-analysis extension of this concept. Students may have trouble recognizing that a fraction with value 1 does not need to have an identical numerator and denominator—they must merely be equivalent. It is important for students to demonstrate the technique on simple problems and then work their way up to harder problems.

As usual, stress organizing work to make the solutions easier to check and easier for another person to verify. Furthermore, keeping the units organized arranges the rest of the problem. As part of the organization process, students need to think about what types of units (mass, volume, time, length) are involved in the problem and what types they want to go toward. Students must also become accustomed to giving appropriate units in their solutions to problems. For example, 37 feet per second would be appropriate for describing the initial velocity of a projectile. On the other hand, the units would be inappropriate to describe the speed of a car near a school zone.

Finding the real-life connections with unit analysis is easy. Much public policy is presented in terms of costs and benefits. One way to look at policy is to examine the cost-to-benefit ratio. For example, if it costs $23,500 per prisoner for one year of incarceration in a maximum-security jail, is there some way to spend $23,500 on preventive action prior to a crime taking place? Some educators argue that because less than half of the general prison population holds a high school diploma or GED (U.S. Department of Justice, 2000), more money spent on educational intervention could increase the graduation rate and reduce the crime rate. In this argument, the costs to government would be reduced and the benefits to society would increase. This issue, as well as other governmental budget issues, can be analyzed and discussed by students.

Another common use of unit analysis is as close as the grocery-store shelf. Most stores feature shelf tags that give the unit price of an item. Note that the shelf tags provide appropriate units: ounces per dollar, ounces per cent, pound per dollar, and so on.

Notes on Text Problems

 **TONI'S TRIP**

Discuss with the class that there is not a unique, correct way to set up units in ratios. Different purposes require different ratios.

CONVERSION PRACTICE

Students need to practice setting up simple conversions and canceling the units before they attempt difficult problems. Although many of these problems can be done by students without setting up the process of canceling the units, emphasize to students that they need to practice this skill on easy problems or they will be lost on the hard problems.

Good Quiz Problems

For a one-problem quiz on this strategy, consider one of the following problems from Problem Set A, Version 2, but with at most five or six parts.

5. Mason's Taxi Service

6. Marcel's Trip

14. Axelrod's Trip

15. Zeke and Friends

Problems like 16, Horse Racing, are good problems for the midterm.

©2004 Key College Publishing, *Instructor Resources: Crossing the River with Dogs*, Johnson/Herr/Kysh

Text Problems

AT THE GROCERY STORE

A 12-ounce can of Rosario's Refried Beans sells for 59 cents, and the 16-ounce size of the same brand sells for 81 cents. Which can is the better buy? Work this problem before continuing.

©2004 Key College Publishing, *Instructor Resources: Crossing the River with Dogs*, Johnson/Herr/Kysh

TONI'S TRIP

Toni drove 80 miles in 2 hours and used 5 gallons of gas. Notice that there are three different types of units of measure here—miles, hours, and gallons—measuring distance, time, and volume, respectively. From this information, you can calculate six ratios of quantities, considering two units at a time. The six ratios are miles/hour, hours/mile, gallons/hour, hours/gallon, gallons/mile, and miles/gallon. Calculate the value of each of these ratios. Work this problem before continuing.

©2004 Key College Publishing, *Instructor Resources: Crossing the River with Dogs*, Johnson/Herr/Kysh

RUNNING FOOTBALL FIELDS

Francisco ran 8 miles. Being a football fan, Francisco wondered how many times he had run the equivalent of a full football field (100 yards). How many times had he? Work this problem before continuing. (Note: 1 mile = 5280 feet)

©2004 Key College Publishing, *Instructor Resources: Crossing the River with Dogs*, Johnson/Herr/Kysh

GAS CONSUMPTION

A car traveling at 65 miles per hour gets 25 miles per gallon and travels for 45 minutes. How many gallons does the car use during that time? Read the solution to this problem, making the manipulatives that the solution suggests and setting up the multiplication needed for the solution.

©2004 Key College Publishing, *Instructor Resources: Crossing the River with Dogs*, Johnson/Herr/Kysh

LEAKY FAUCET, PART I

A leaky faucet drips 1 fluid ounce every 30 seconds. How many gallons of water will leak from this faucet in 1 year? Solve this problem before continuing.

©2004 Key College Publishing, *Instructor Resources: Crossing the River with Dogs*, Johnson/Herr/Kysh

©2004 Key College Publishing, *Instructor Resources: Crossing the River with Dogs*, Johnson/Herr/Kysh

Text Problems (continued)

CONVERTING WOOD

Raoul had done all the measurements perfectly. He needed a piece of plywood that measured 122 centimeters by 244 centimeters. The problem was, with all of his traveling, he had gotten used to the notion that if it's Tuesday, this must be Belgium. Unfortunately, it was Thursday and he had jetted to the United States the night before. What size piece of wood should Raoul ask for at the local hardware store, where they don't use the metric system? Solve this problem before continuing.

©2004 Key College Publishing, *Instructor Resources: Crossing the River with Dogs,* Johnson/Herr/Kysh

CONVERSION PRACTICE

Work the following problems. You may use any English-to-English conversions and any metric-to-metric conversions you want. However, the only English-to-metric conversion you may use is 1 meter = 3.281 feet.

1. Change 75 kilometers to miles.

2. Change 18 inches to centimeters.

Work these problems before continuing. Round answers to the nearest hundredth.

©2004 Key College Publishing, *Instructor Resources: Crossing the River with Dogs,* Johnson/Herr/Kysh

FASTBALL

Randy Johnson, currently of the Arizona Diamondbacks, has been clocked throwing a baseball 100 miles per hour. At that speed, how much time does the batter have to react? (That is, how much time before the ball reaches the plate?) The pitcher's mound is 60 feet 6 inches from home plate. Work this problem before continuing.

©2004 Key College Publishing, *Instructor Resources: Crossing the River with Dogs,* Johnson/Herr/Kysh

AREA

Find the area of a rectangle with length 3 feet and width 2 feet. Solve this problem before continuing.

©2004 Key College Publishing, *Instructor Resources: Crossing the River with Dogs,* Johnson/Herr/Kysh

©2004 Key College Publishing, *Instructor Resources: Crossing the River with Dogs,* Johnson/Herr/Kysh

Text Problems (continued)

LONG COMMUTE

Gerónimo and three friends regularly drive 208 miles. Their car gets 35 miles per gallon. Gas costs $1.59⁹/10 per gallon. They drive at an average speed of 50 miles per hour. Find each of the following:

1. Gallons of gas used

2. Hours the trip took

3. Average feet per second

4. Cents per passenger-mile

Solve this problem before continuing.

MR. ROGERS' NEIGHBORHOOD

Janice, Shane, Rose, and Gina are going to be paid $84.70 for cleaning up Mr. Rogers' neighborhood. They each worked 5 hours, except Rose, who was 45 minutes late. How much should each be paid? Work this problem before continuing.

Problem Set A, Version 2

1. MORE CONVERSIONS

Convert from metric to English or from English to metric as indicated. The only metric-to-English conversions you are allowed to use are

1 m = 3.281 ft

1 gal = 3.79 L

Of course, you may use any English-to-English conversions (such as 1 mi = 5280 ft) and any metric-to-metric conversions (such as 1 km = 1000 m).

a. 45 m to feet

b. 280 ft to meters

c. 250 mi to kilometers

d. 31 km to miles

e. 35 ft to centimeters

f. 24 cm to inches

g. 48 in. to millimeters

h. 50 mi/hr to $\dfrac{\text{meters}}{\text{second}}$

i. 3 gal to liters

j. 20 L to quarts

2. LIQUID PIPELINE

Make a pipeline diagram like that described in Chapter 8 for the following liquid measures: pint, quart, gallon, liter, centiliter, and milliliter.

3. TIME IN A PIPELINE

Make a pipeline diagram like that described in Chapter 8 for the following time measurements: second, minute, hour, day, week, month, year, decade, and century.

4. WHAT'S UP, DOC?

A bag of carrots costs $1.75 and has 25 carrots in it. Since the bag is 5 lb, find each of the following:

a. dollars per carrot

b. cents per carrot

c. ounces per carrot

d. carrots per pound

©2004 Key College Publishing, *Instructor Resources: Crossing the River with Dogs,* Johnson/Herr/Kysh

Problem Set A, Version 2 *(continued)*

5. **MASON'S TAXI SERVICE**

Mason flew his plane 550 miles in 3 hours 40 minutes. He used 9 gallons per hour, and the fuel cost him $1.85 per gallon. He took three passengers with him. Find each of the following:

a. gal

b. mi/hr

c. ft/sec

d. dollars/hr

e. mi/gal

f. dollars/passenger

g. cents/mi

h. total number of passenger-miles

i. passenger-miles/gal

j. cents/passenger-mile

6. **MARCEL'S TRIP**

Marcel and five friends drove at a speed of 60 miles per hour over a distance of 420 miles. The trip cost $16.24 in gasoline and they used 2 gal/hr. Find each of the following:

a. gal

b. hr

c. ft/sec

d. dollars/hr

e. dollars/gal

f. dollars/passenger

g. cents/mi

h. total number of passenger-miles

i. passenger-miles/gal

j. cents/passenger-mile

7. **COMMUTER FLIGHT**

An airplane is traveling 550 miles per hour. It is carrying 80 passengers. Answer the following questions:

a. How many miles per minute?

b. How many feet per second?

c. How many passenger-miles on a 400-mile trip?

d. How many passenger-miles per hour?

e. How many passenger-miles per minute?

8. **A SHOT IN THE DARK**

A projectile is shot into space at 60 meters per second.

a. Determine its speed in miles per hour.

b. Determine its speed in kilometers per hour.

TEACHING RESOURCES

©2004 Key College Publishing, *Instructor Resources: Crossing the River with Dogs*, Johnson/Herr/Kysh

Problem Set A, Version 2 *(continued)*

9. IT'S ABOUT TIME

Mrs. Ralls hired Cinnamon, Rickell, Jeannette, and Kari to clean her garage. They each worked 6¼ hours, except Jeannette, who started early and worked 7½ hours. They were paid a total of $126. Each worker will be paid at the same rate. How much should each person get?

10. AFTER THE DEADHEADS

Sassafras, Moonshine, Peace, Harmony, and Chynna were hired to clean up after a Grateful Dead concert. Moonshine started very late but worked the 3 hours until dawn. Sassafras put in 5⅔ hours, and Harmony worked as much as Sassafras but also got in an extra third of an hour. Peace had never worked before but was hired to work 4 hours because somebody decided to give Peace a chance. Chynna worked 8 hours because she was isolated for the longest time and refused help from anybody else. The five were paid $128.80 total. If each person was paid the same hourly rate, how much did each person earn?

11a. WHERE'S THE RUE?

Tova used to spend 40 minutes per day jogging and covered a distance equivalent to the distance from Rue de Vache to LaPointe 73 times during the year. The distance is 20 miles. What is her speed in miles per hour?

11b. WHERE'S THE VACHE?

Tova now spends 45 minutes per day jogging and has sped up 1 mile per hour. How many times would she run a distance equivalent to the distance from Rue de Vache to LaPointe in a year?

12. MANUSCRIPT

A manuscript page generally has about 500 words on it. If you read a manuscript page in 1 minute 24 seconds, what is your reading rate in words per minute? What is the number of pages you read per hour?

©2004 Key College Publishing, *Instructor Resources: Crossing the River with Dogs*, Johnson/Herr/Kysh

Problem Set A, Version 2 *(continued)*

13. EVENING EXERCISE

Sylvia walks 4 miles every evening. She is also a golfer. Since playing an average 18-hole golf course involves walking 6300 yards total, she wondered how many holes she walked the equivalent of every time she went for her evening walk. (By the way, don't worry about Sylvia. She hits the golf ball straight, so there's no walking from one side of the course to the other side when she plays golf.) What is the equivalent number of holes Sylvia walked each evening?

14. AXELROD'S TRIP

Axelrod drove 396 miles on 14.4 gallons of gas. It took him 7 hours 30 minutes. Gas for his car costs $1.30 per gallon. Find the following:

   a. miles per hour

   b. total fuel cost

   c. miles per gallon

   d. feet per second

   e. cents per mile

15. ZEKE AND FRIENDS

Zeke and three friends drove for 300 minutes, using 8 gallons of gas. They got 30.5 miles per gallon with a total fuel cost of $9.84. Find the following:

   a. miles

   b. dollars per gallon

   c. miles per hour

   d. passenger-miles per gallon

   e. cents per passenger-mile

16. HORSE RACING

If the horse Seabiscuit travels at a constant speed of 40 miles per hour, how many seconds will it take Seabiscuit to run a race of 6 furlongs? (Note: 1 furlong = $1/8$ mile)

TEACHING RESOURCES

©2004 Key College Publishing, *Instructor Resources: Crossing the River with Dogs*, Johnson/Herr/Kysh

Problem Set B, Version 2

1. **I CAN SEE FOR MILES AND MILES**

Last week I drove north on Interstate 77 to Cleveland. There was a speedometer check along the side of the road with posted signs 1 mile apart. I timed us on my wristwatch and found that it took us 4 minutes 30 seconds to travel 4 miles. We had the car set on cruise control, so we were traveling at a constant speed. A while later we passed a sign that said "Akron 90 miles." How long did it take us to drive those 90 miles to Akron? Answer in hours, minutes, and seconds. (For example: 7 hours 6 minutes 15 seconds.)

2. **AN AGE-OLD PROBLEM**

When Lee was thrice as old as Kevin,
His sister Kate was twenty-seven.
When Kevin was half as old as Kate,
Then brother Lee was thirty-eight.
Their ages add to one forty-three.
How old are Kevin, Kate, and Lee?

3. **CROSS-NUMBER PUZZLE ONE**

Find the two solutions to this cross-number puzzle. All numbers are three-digit numbers, entered one digit per box. None of the three-digit numbers are the same. It is possible to have repeated digits in a number. There are no zeros in the puzzle.

ACROSS

1. The last two digits form a two-digit number that is a power of the first digit.

4. A square

5. A cube

DOWN

1. A Fibonacci number

2. All digits are even.

3. A prime number

1	2	3
4		
5		

Note: The Fibonacci sequence (discussed in Chapter 5: Look for a Pattern) is the sequence 1, 1, 2, 3, 5, 8, 13, 21, ...

©2004 Key College Publishing, *Instructor Resources: Crossing the River with Dogs*, Johnson/Herr/Kysh

Problem Set B, Version 2 *(continued)*

4. FILLING THE PLANTER BOX

A home builder built a large planter box in the backyard. The planter box is 20 feet long, 3 feet back, and 2 feet high. The builder wants to fill the box with dirt. He is able to buy dirt in 3-pound bags. Dirt weighs 2 pounds per cubic foot. He has a wheelbarrow that will carry three bags of dirt at a time from his garage (where his truck with the dirt is) to the backyard. How many trips does he have to make with the wheelbarrow to fill up the planter box?

5. GENEROUS FRIENDS

Two friends went out for pizza and soda after work one day. They each ordered and drank one large soda and ate part of their pizza. They then decided they were still thirsty, so Molly went up to get another soda. Nancy didn't have any money, so she asked if she could share some of Molly's soda. Molly then poured half of her soda into Nancy's glass. Nancy protested, saying it was too much, and she poured one-third of the soda in her glass back into Molly's glass. Molly insisted that Nancy have some more, so she poured one-fourth of the soda in her glass back into Nancy's glass. Of course, Nancy would not take so much soda, so she poured one-fifth of the soda in her glass back into Molly's glass. This went on for quite a while, with each pouring increasing the denominator of the fraction by 1. After 99 pourings, they decided to stop. At this point, how full was each glass of soda?

TEACHING RESOURCES

©2004 Key College Publishing, *Instructor Resources: Crossing the River with Dogs,* Johnson/Herr/Kysh

Problem Set B, Version 3

1. DISCOUNT HARDWARE

A carpenter went to a discount hardware store to buy hammers, screwdrivers, and large nails. Hammers cost $10 each, screwdrivers cost $7 each, and large nails cost $0.50 each. The carpenter bought exactly 100 items and spent $150. How many of each did she buy?

2. LIFE SPAN

Assume that a human heart is capable of beating about 2.8 billion times in a lifetime. If a person's average heart rate is 72 beats per minute, what is that person's life span? Express your answer in years and days. (For example: 5 years 142 days.)

3. COIN COLLECTION

Trevor has a piggy bank full of coins. One day he counted them and said that he had an equal number of pennies, nickels, dimes, and quarters. His father, Eric, handed him three coins. Trevor now had $7.22. What were the three coins that his father gave him?

4. THE GOAT PROBLEM

A goat is tethered to one corner of a 20-foot-by-35-foot barn in the middle of a large grassy field. The rope is 50 feet long. Over what area can the goat graze? (He can't go inside the barn, and there isn't any grass in there anyway.)

5. **CROSS-SUMS**

Cross-sums puzzles are very popular in crossword magazines. The rules are as follows: The numbers shown in the black squares are the sums of the digits that you will fill into the white spaces, one digit per space. A number above the diagonal is the sum of the digits in the white spaces to the right. The number below the diagonal is the sum of the digits in the white spaces below. Important rules: No zeros are used, and a digit cannot appear more than once in any digit combination. For example, if the sum of a two-digit combination is 6, then the two digits can be 15, 24, 42, or 51, but not 60 (no zeros allowed) or 33 (no repeated digits allowed).

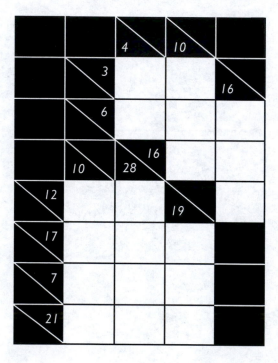

Problem Set B, Version 4

1. WALLY WEIRD'S WINE

Wally, the owner of Weird's Wine Company, had a strange way of deciding in which years he produced new wine. He would produce a new wine in any year in which the sum of the digits in the year the wine is produced is equal to the sum of the digits in the year the wine is sold is equal to the age in years of the wine when it is sold. In what years in the twentieth century did the company produce wine?

Wally's granddaughter, Wanda, took over the company in the year 2000. She wished to keep up Wally's tradition. In what years in the twenty-first century will the company produce wine?

2. LAKE MEAD FLOODS THE MOJAVE DESERT

The lake created by Hoover Dam on the Arizona-Nevada border has a capacity of 31,250,000 acre-feet. If the whole lake suddenly flooded the Mojave Desert (area 15,000 square miles), how deep would the water be? Assume the Mojave Desert is completely flat. (An acre-foot is the volume of water that would cover an area of 1 acre to a depth of 1 foot. An acre is 43,560 square feet.)

3. CROSS-NUMBER PUZZLE TWO[1]

The puzzle below is a cross-number puzzle, similar to a crossword puzzle except that the entries are numbers. Enter one digit per square. The thick, heavy line is a separator.

ACROSS

a. A prime number

c. The sum of the digits of *a* across

DOWN

a. Square of the sum of the digits of *b* down

b. A prime number

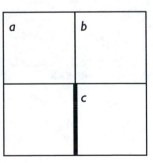

---

[1]This problem was written by Luther Burbank High School student Erik Maness.

Problem Set B, Version 4 *(continued)*

4. VCR

A VHS tape for a videocassette recorder can record 2 hours of programming at the SP speed, 4 hours of programming at the LP speed, and 6 hours of programming at the EP speed. The quality of the recording goes down as the time gets longer (the quality of SP is much better than the quality of LP, which is in turn better than the quality of EP).

Felicia has a VHS tape that has recorded the following TV shows:

Sesame Street: 1 hour recorded at LP

Mr. Rogers' Neighborhood: 30 minutes recorded at EP

Reading Rainbow: 15 minutes recorded at SP speed

Pinnochio: 40 minutes recorded at LP speed

Charlie Brown: 10 minutes recorded at SP speed

Part 1: She wants to record *Rudolph the Red-Nosed Reindeer* (which is a 1-hour show) on this tape without erasing any of the other shows. She would also like to have the highest quality recording possible, given the amount of tape she has left. On which speed should she set her VCR and how much time (at that speed) will she have left on the tape after recording this show?

Part 2: Suppose she wanted to optimize quality by recording as much of *Rudolph* as possible on SP and then switching to LP at just the right time to record the entire show and use up the whole tape. At what moment should she switch from SP to LP to do this?

5. SPRINKLERS

At each corner of a 16-ft-by-16-ft square lawn there is a sprinkler that waters one-quarter of a circle of 8-foot radius. In the center of the lawn, there is another sprinkler that waters a full circle of 8-foot radius. Part of the lawn is being watered by two sprinklers, and part of the lawn is being watered by only one sprinkler. What percentage of the lawn is being watered by two sprinklers?

9 Solve an Easier Related Problem

W̶hen a problem is hard to solve, why not just solve an easier one? An easier related problem (ERP) needs to be (1) easier, (2) related, and (3) a problem. This strategy provides a way of changing focus, because instead of looking at the original problem, you look at a new problem that you create, based on the original. This strategy also has elements of organizing information.

Easier related problems are analogous to guess-and-check. With the guess-and-check strategy, you try a number to see if it works. With ERPs you try a process to see if it works. This can be done in a number of ways. Sometimes when using an ERP, you end up trying specific numbers and looking for a pattern. All of these methods are demonstrated in the chapter.

Below is Tom Sallee's list of the various ways to make a problem simpler. The problems listed use the methods shown in italics and are from either the text or from Problem Set A.

1. *Use a number instead of a variable.* We used this in Averages and in Square and Hexagon.

2. *Use a small or easier number in place of a more difficult one to develop the process for solving the problem.* We used this in Simpletown Elections, TV Truck, Exponents and the fraction-to-decimal example, Averages, and Potatoes.

3. *Do a set of specific easier examples and look for a pattern.* We used this in From One to One Hundred, How Many Squares?, Fifty-Two-Card Pickup, Last Digit, Diagonals, Sum of Odds, Chinese New Year, and Remainder.

4. *Do a specific easier example and figure out an easier process that will work to solve the problem.* We used this in Divisors and Reciprocals and in Odd and Even.

5. *Change, fix, or get rid of some conditions.* We used this in Good Luck Goats, Next Train East, Inscribed Square, Twenty-Five-Man Roster, and Fifty-Two-Card Pickup.

6. *Eliminate unnecessary information.* We used this in Simpletown Elections and in Potatoes.

This strategy can be very difficult to master. The key question to ask is "How can I make this problem easier?" Students must learn not only how to apply the strategy, but also to look for instances where it is appropriate. If a problem seems too hard or too confusing, they should look for an ERP. Sometimes you'll need more than one ERP to solve the original problem.

Sometimes irrelevant information is written into a problem. An ERP for such a problem would exclude the irrelevant information.

The idea of using a simpler problem in learning mathematics can be very powerful. Knowledge of some facts, such as $\frac{1}{2} = 0.5$, can allow a student to extend the concept to the general process of changing a fraction to a decimal. For an easier problem to work, the answer to the easier problem must be within the student's grasp.

Estimation also plays a role in ERPs, but it is different from the strategy. The purpose of an ERP is to allow you to determine the process needed to solve the problem. On the other hand, an estimate can be done only when you know the process. Estimation allows you to verify that you have successfully performed the process.

ERPs in real life can be well hidden. The strategy is powerful enough that it extends a great distance outside the realm of mathematics. A nurse, Allyson, explained that she uses ERPs when she needs to stick an intravenous needle into a patient's forearm. Sometimes, usually due to advanced age or poor health, a patient has veins that are very difficult to locate. So Allyson looks at her own arm—she is in good health and is young; both factors are helpful in locating veins. She finds a good blood vessel and then searches for other physical landmarks around it on her arm. That is the easier problem. She then takes that knowledge back to the original (and difficult) problem of finding the blood vessel on the patient. She uses the corresponding physical landmarks on her patient to determine where to attempt to stick the needle.

Notes on Text Problems

SIMPLETOWN ELECTIONS

This is a case where organizing the information comes in very handy. Students also need to use simplified numbers and ignore the irrelevant information.

HOW MANY SQUARES?

The process in this solution is to start with the easiest problem (a one-by-one square), do easier problems, organize the data, and then develop a pattern.

Notes on Text Problems (continued)

DIVISORS AND RECIPROCALS

The process in this problem is to pick a small number so that the divisors are easy to add up. The prediction is then tested on another example. The same kind of pattern shows up again. Thus, it is very likely, and it makes intuitive sense, that this pattern applies to the original problem.

GOOD LUCK GOATS

For a good article about this type of problem, see "The Riddle of the Vanishing Camel" by Ian Stewart in the Mathematical Recreations column of the June 1992 issue of *Scientific American*. This article can be found online at www.cthisspace.com/ftl/macaw/MR49.html.

AVERAGES

The given averages and the number of quiz scores in this problem are difficult numbers. We use two types of easier related problems to solve this problem: (1) We replace a variable with a number to see what is going on, and (2) we use easier numbers. Both of these substitutions make the problem much more manageable and give us a plan of attack. From that point on, it is just a matter of applying to the hard problem the procedure learned on the easier problem.

INSCRIBED SQUARE

This problem can be made easier by eliminating one of the conditions. Students will also need to draw lots of diagrams.

Notice that a lot of these problems can probably be done by programming a computer. Technology can therefore nullify cleverness to a certain extent. This does not mean that learning these strategies is a dead end on Technology Avenue; rather, the thinking skills developed, the flexibility learned, and the specific strategies will be more important, as those are areas of mathematics that have yet to be conquered by computers.

Good Quiz Problems

We recommend no quiz on this chapter. This strategy is included as part of the suggested midterm.

Text Problems

FROM ONE TO ONE HUNDRED

What is the sum of the first 100 whole numbers? Work this problem before continuing.

■ **SIMPLETOWN ELECTIONS**

The clerk of Simpletown has the job of getting materials ready for the next municipal election. There are 29 issues and candidates. In the last election, there were 28,311 registered voters, representing 18,954 households, and they voted at 14 polling places. (In Simpletown all registered voters vote in the election.) The clerk figures she needs a proportionate amount of materials for this election. This time there are 34,892 people registered to vote. How many polling places will be needed? Work this problem before continuing.

■ **HOW MANY SQUARES?**

How many squares are there on a checkerboard? (*Hint:* It is more than 64.) Work this problem before continuing.

■ **DIVISORS AND RECIPROCALS**

The divisors of 360 add up to 1170. What is the sum of the reciprocals of the divisors of 360? Work this problem before continuing.

■ **EXPONENTS**

Simplify each expression:

$$m^{1/8} \cdot m^{2/3} \qquad \left(y^{1/3}\right)^{6/7}$$

Work these problems before continuing.

Text Problems (continued)

GOOD LUCK GOATS

In the mythical land of Kantanu, it was considered good luck to own goats. Barsanta owned some goats at the time of her death and willed them to her children. To her first born, she willed one-half of her goats. To her second born, she willed one-third of her goats. And last she gave one-ninth of her goats to her third born.

As it turned out, when Barsanta died she had 17 goats. Barring a Solomonic approach, how should the goats be divided?

Work this famous problem before continuing.

AVERAGES

The average of a group of quiz scores is 31.8. There are k quiz scores in the group. The average of 10 of these quiz scores is 24.3. Find the average of the remaining quiz scores in terms of k. Work this problem before continuing.

NEXT TRAIN EAST

A train leaves Roseville heading east at 6:00 a.m. at 40 miles per hour. Another eastbound train leaves on a parallel track at 7:00 a.m. at 50 miles per hour. What time will it be when the two trains are the same distance away from Roseville? Work this problem before continuing.

INSCRIBED SQUARE

Given any **triangle,** draw a square inside of it so that all four vertices of the square are on the sides of the triangle. Two of the **vertices** of the square should be on one side of the triangle, and each of the other two sides of the triangle should have one vertex of the square.

Make the problem easier by eliminating one of the conditions. You'll need to draw lots of diagrams. Work this problem before continuing.

Problem Set A, Version 2

1. LAST DIGIT AGAIN

What is the last digit in the product of $(3^1)(3^2)(3^3)(3^4) \ldots (3^{398})(3^{399})(3^{400})$?

2. DIAGONALS OF A POLYGON

A certain convex polygon has 14 sides. How many diagonals can be drawn?

3. SUM NUMBERS—LOTSA NUMBERS

Find the sum of the first 8000 multiples of 3.

4. AIR FARE

Marlene is responsible for making sure that enough packaged meals are put on several flights for her company. She needs to get 86 meals on Flight 914, which is leaving at 10:23 a.m. from Gate 27 at Metro Airport. They usually fly a 747 on that route, though sometimes they use a 737 when less than 92 passengers are ticketed. She also needs to make sure that she gets 134 meals, including 9 vegetarian meals, loaded onto Flight 797, which leaves at 12:13 p.m. The flight takes 4 hours 22 minutes. She and the other 28 employees of ACME Airmeals have a total of 19 flights that they are responsible for. What is the difference in the departure times for the two flights mentioned?

5. SEASON TICKET PLANS

The Carolina Shores baseball team considered offering a unique season ticket plan. You could choose anywhere from 1 game to 42 games (all are home games). How many different ticket plans does this really mean they have? (How many different combinations of home games are there?)

6. ONCE A YANKEES FAN, ALWAYS A YANKEES FAN

My younger sister Gianna is a Yankees fan: always has been and, the way I see it, always will be. When she was about 2 and I was just a little older, she used to tell me who her favorite players were. She'd pick up the stack of Yankee trading cards and show me her favorite player. Sometimes it was Mickey Mantle. Sometimes it was Roger Maris. Sometimes she had two or three favorites, like Joe Pepitone, Tony Kubek, and Clete Boyer. Once she went through the whole stack, saying, "This is my favorite, this is my favorite—all 25 of them!" In how many different ways can Gianna select from 1 to 25 of her favorite players out of the stack of 25 baseball cards?

Problem Set A, Version 2 *(continued)*

7. **CLASSIC MUSTANGS**

None of us really got along as friends, but we shared several things in common. We had some joint business dealings and a love of old Mustangs. We only liked the ones from 1964½ through the 1968 model year. When we found out that the collection of Mustangs from the Sparks Auto Museum was going to be auctioned off, we put our minds and our money together. Instead of bidding against each other, we decided to bid as a group on each Mustang available. We set a maximum price we would bid, and quite frankly, none of us had any favorites; we loved them all equally. We also agreed to divide "our take" as follows: ½ to Travis because he was usually the big money on our projects, ¼ to Sandra as she was usually the brains that got things done, and I was to get a ⅙ share. In the end we were the winning bidders on 11 Mustangs. How many Mustangs do each of us get?

8. **VIVE LA DIFFERENCE!**

Find the difference between the sum of the first 500 multiples of 3 and the sum of the first 500 odd numbers.

9. **SAVING PENNIES**

Ricardo kept saving pennies. Every day he saved the same number of pennies as the day's date (for example, 12 pennies on March 12). Maritza did something different. She saved 5 cents on the first day of the month. She then saved 5 cents more each day than she had the previous day. For the month of March, who had saved more money, and how much more money was it?

10. **BUSINESS CARDS**

Florin opened up a café. On the first day, he had no customers. On the second day, however, he had 5 customers. On the third day, there were 10 customers, and on the fourth day there were 15 customers. He also ran a lunch giveaway, whereby if you left a business card, he would enter it in a drawing for a free lunch. On the first day, no one left a card (since there were no customers); on the second day, 3 people left cards; and each following day, 3 more people left business cards than on the previous day. If this pattern continues for a full year (365 days), what is the difference between the total number of customers he would have and the total number of business cards?

©2004 Key College Publishing, *Instructor Resources: Crossing the River with Dogs*, Johnson/Herr/Kysh

Problem Set B, Version 2

1. COVERING THE PATIO

Mr. Smith needs to nail some fiberglass pieces to a wooden frame to cover the patio on the back of his house. The patio measures 30 feet by 20 feet, and each fiberglass piece measures 2 feet by 6 feet. (The panels are actually slightly bigger to allow for overlapping.) He needs to nail each panel to a wooden frame (which is the same size as the patio), placing the nails along each of the four edges of the fiberglass. The nails will be spaced 1 foot apart. How many nails does he need to put all of the fiberglass in place? (*Note:* A nail can be driven into several pieces of overlapping fiberglass.)

2. LIKE A WILDFIRE

The volunteer firefighters were frustrated with their efforts at basic fire-prevention training. It seemed that no matter how many fire-safety meetings they set up, hardly anybody ever paid attention. They devised a plan. They decided that each of them (there were eight firefighters) would teach two other people the fire-safety basics. At that point, the teacher would retire but each student would then teach two others. Those people, in turn, would teach two others. The whole thing would be mandated by the city council, and each person would have a month to fulfill his or her teaching requirement. The firefighters taught the first group of people in the first month. Under this plan, how many people would know the fire-safety basics after 10 months?

3. SUM OF TEN

If you add the digits in a number, how many numbers between 0 and 10,000 will have a sum of 10? (For example, 334 is one such number because $3 + 3 + 4 = 10$.)

4. LAPPING JOGGERS

Jan can run around a quarter-mile track in 90 seconds. Silvia can run around the same track in 72 seconds. They started running in the same direction from the same place at the same time. The two women agreed to stop when Silvia caught up to Jan for the third time (lapping her twice, then catching up to her again). How far did each woman run?

TEACHING RESOURCES

©2004 Key College Publishing, *Instructor Resources: Crossing the River with Dogs*, Johnson/Herr/Kysh

Problem Set B, Version 2 *(continued)*

5. **THE PHOON BROTHERS**

The Phoon brothers—Buff, Ty, and Lam—drove from Lane County, Oregon, to Disneyland, a total of 1020 miles. They picked up Helen Highwater in Stockton, which is 340 miles from Disneyland. Their car got 30 miles per gallon. Gas cost $1.15 per gallon. How much should each pay? (*Note:* There are at least two different ways to do this, and the answers come out quite different. Be prepared to defend your result.)

Problem Set B, Version 3

1. PYRAMID SCHEME

Aldo and his two buddies decided to run a pyramid investment scheme. The plan was simple. They told investors that if they paid $1,000, in one week the investors would receive $1,500. On their part, the investors had to also bring two more people into the pyramid that week. The new people would need to pay $1,000 and bring in two new people the next week. Aldo and his buddies each started by bringing in two people as their first set of investors. All went well for the first 12 sets of investors, but then it all fell apart and Aldo and his buddies were arrested. How many people (other than the three buddies) had invested by the time the pyramid was flattened? (*Note:* Pyramid schemes are illegal and perpetrators are dealt with severely by the courts.)

2. LOSING TIME

When I reset my clocks for daylight saving time, I took care to make sure that each one had exactly the same time on it. A week later, however, I noticed that the clock on the VCR was 14 minutes slow. I didn't have time to reset it, and as I drove to work, I noticed that my car clock was 7 minutes fast. I decided not to reset the slow clocks, but I finally got sick of it and changed them when the car clock was exactly an hour ahead of my VCR clock. I initially set the clocks to daylight saving time on a Sunday. How many days later, and on what day of the week, did I reset them?

3. DRIVE ME WILDE

The Wilde cousins—Fawn, Kat, and Wolfgang—decided to drive to Walt Disney World, a 630-mile drive from their home. One hundred and ten miles into the drive, they picked up their cousin, Duckie. In Florida, about 115 miles from their destination, they picked up another cousin, Mildred. (If you're wondering about her name, it's different because she's not from the Wilde side.) Their car got 35 miles per gallon and gas cost them $1.13 per gallon. They all agreed to share the gas costs. How much should each cousin pay? (*Warning:* There are two valid ways to compute the shared costs. Be prepared to defend your answer.)

Problem Set B, Version 3 *(continued)*

4. SOCCER LEAGUE

In the Great Falls, Minnesota, soccer league, there were six teams:
Allosaurus, Brontosaurus, Cetiosaurus, Dimetrodon, Hypsilophodon,
and Triceratops. (The kids in the soccer league were really fond of
dinosaurs.) Each team played each other team twice. At the end of the
season, the sports editor of the *Great Falls Gazette* was looking over her
notes from the season before publishing the results in the paper. Her
son played on Brontosaurus and her niece played on Dimetrodon. She
remembered that no team tied more than one game. She also had
written down the number of play-off points that five of the teams had
acquired. (A play-off point is awarded as follows: 2 points for a win,
1 point for a tie, 0 points for a loss.) Help her determine the final
standings. She also wanted to know who tied whom and how her
son's team did in the two games against her niece's team.

PLAY-OFFS				
Team	Wins	Losses	Ties	Play-off points
Allosaurus				3
Brontosaurus				10
Cetiosaurus				20
Dimetrodon				11
Hypsilophodon				
Triceratops				1

5. RUNNING ERRANDS

Willy has ten errands to run. However, he has his three kids with him,
who are all very young. With all of the waiting in stores, getting in and
out of car seats, naps, diapers, potty breaks, whining, crying, and so
forth, Willy sticks to a rule of thumb: no more than three errands at a
time before going home. How many different sets of three errands can
he arrange?

Problem Set B, Version 4: Gary's Van

I. TWO VEHICLES

Gary has two cars: his fairly new van and his old clunker station wagon. The wagon currently has 16 times as many miles on it as the van had when the wagon had 3 times as many miles as the van has now. Between now and then, each vehicle has been driven the exact same number of miles. The wagon has not yet been driven 100,000 miles. All mileage amounts are in thousands. How many miles has the van been driven now?

2. DIGITAL CLOCK

In Gary's van, the digital clock is not working right. One or two of the lines in each digit are always shorting out, although not necessarily the same lines in each digit and not the same lines all the time. One day, Gary drove over to his friend Jeff's house. When Gary got into the van, the clock read the time shown below left. When he got to Jeff's house, a drive of no more than a half hour, the clock read the time shown below right. What time was it when he got to Jeff's house?

   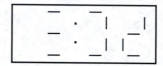

3. WATCH YOUR SPEED

Gary had to buy a watch because the clock in his van was so unreliable. Unfortunately, he bought a used watch and it didn't work very well. He compared it to his good clocks in his house and found that his watch was fast: When his watch showed 1 minute had passed, it had really been only 57 seconds. One day, Gary was taking a trip to Monterey. He suspected that the speedometer in his van was off, but he wasn't sure by how much. He found one of those mileage checks on the freeway. He drove 5 miles (according to the mileage signs, which you may assume are accurate) in 5 minutes 30 seconds according to his watch. He kept his speedometer on 53 miles per hour the whole time. If the speed limit is 65 miles per hour, what is the fastest he can drive according to his speedometer and avoid breaking the law? (*Note:* His speedometer reads 0 miles per hour when he is stopped.)

TEACHING RESOURCES

©2004 Key College Publishing, *Instructor Resources: Crossing the River with Dogs,* Johnson/Herr/Kysh

Problem Set B, Version 4: Gary's Van (continued)

4. TWO-LANE HIGHWAY

On Gary's trip to Monterey he found himself in the midst of 100 cars, driving on what seemed to be an endless two-lane highway (one lane in each direction). The cars were all closely bunched because they were traveling in reverse order of their desired speeds. In other words, the person who was in the lead car wanted to travel the slowest and thus had the huge line of cars behind him. The person following the lead car wanted to go faster than the lead car but slower than everyone else. The person following the first two cars wanted to go faster than the first two cars, but slower than everyone else. This pattern continued, with each person wanting to go faster than everyone in front of them but slower than everyone behind them. The drivers all decided that they wanted to pass everyone in front of them and be passed by everyone behind them so that the entire order of 100 cars would be reversed.

If more than one passing can take place at the same time, what is the least number of passing periods that must occur for all the cars to end up in the desired order?

(*Note:* Only two cars can be involved in each passing—the passer and the passee—but multiple passings can occur simultaneously. For example, car 2 could pass car 1 at the same time that car 4 passes car 3, at the same time that car 52 passes car 51, and so on. All of that would only count as one passing period.)

5. PEDESTRIAN

Lori was walking down the street on her way to the grocery store. As she got to a crosswalk, she noticed a van coming from her left, passing under a railroad overpass just as she started to cross the street. As she finished crossing the street, the van reached the crosswalk behind her. (It was Gary, coming home from Monterey.) She turned left when she reached the sidewalk. It took her 15 seconds to cross the street and 2 minutes 20 seconds to reach the overpass. She walks 3 miles per hour. How fast was the van going? (Assume it was traveling at a constant speed.)

©2004 Key College Publishing, *Instructor Resources: Crossing the River with Dogs*, Johnson/Herr/Kysh

10 Create a Physical Representation

I f a picture is worth a thousand words, then a model is worth a million. *Physical representation* is an umbrella term that encompasses strategies that could be considered separately. It includes using manipulatives, making models, and acting out problems. It is sometimes difficult to determine where each of these strategies ends and another begins.

Elements of guess-and-check are also involved in physical representations. Much of solving a problem with manipulatives involves making incorrect guesses and adjusting subsequent guesses. An advantage of manipulatives in doing guess-and-check is that it is possible to guess, check, and revise very quickly. Using manipulatives also differs from guess-and-check in that you do not usually keep a record of incorrect guesses with manipulatives, whereas in paper-and-pencil guess-and-check, the record is an integral element of the strategy.

The different variations of physical representations have distinguishing characteristics. Models, for example, generally need to have some critical elements made to scale and need to function in the same manner as in the items they represent. That is not necessarily true of a manipulative, which might have no scaled elements. A manipulative is generally set up to deal with spatial relationships and may or may not be a realistic rendition of the item it represents.

Models can physically represent such mathematical concepts as magnitude, scale, quantity, directional movement, positional relationship, static relationship, number, number combinations, relative size, or orientation. Any useful model represents at least one key element. In many cases, models are extensions of diagrams.

While much of the power of modern mathematics comes from abstraction, we risk overlooking real-world connections when we rely too heavily on abstract methods for solving problems. As we learn more and more abstract methods, we may even lose our capacity for using models and manipulatives in problem solving. Fortunately, current trends in mathematics education toward real-world applications should prevent teachers and students from losing their facility with physical representations.

Using models and manipulatives can be especially helpful for students who are kinesthetic learners. It gives other students the opportunity to develop skills they might not otherwise have developed.

When you teach this section, you should recognize a number of things. Many of us did not have strong role models for instruction with manipulatives. We are further disadvantaged in that many of our students will enter our classes with similarly scant experience. Concepts that may seem obvious to us may need to be emphasized to students.

Students may also enter the class with a negative mind-set against manipulatives. They may develop this attitude from their lack of facility and familiarity with the strategy, or they may learn it directly from teachers or people at home with the same attitude. ("We didn't do it that way, and I learned this just fine.")

Finally, people sometimes avoid manipulatives because of the time it takes to set them up.

You will need to spend time emphasizing the benefits of physical representations, such as how they allow you to check guesses easily. Point out how the problems in the text are more easily solved with manipulatives. Students may also be motivated by how physical representations connect with other disciplines. There are many real-life applications of physical representations that you can relate to your students. Some examples are mechanical or civil engineering projects, planning processes (the volleyball examples in the text are from real recreational-league situations), and logistical operations (such as a war room). The article about the Hubble Space Telescope, reprinted in the text, is a good example of a real-life application of manipulatives.

A manipulative approach should not be used for its own sake. There should be underlying purposes: making the problem easier, approaching it through another learning modality, bringing out key elements, teaching the use of manipulatives directly, and so on. An optional section on converting problems to a manipulative approach is included for this chapter.

Using models will improve your students' spatial visualization skills. They will gain concrete experience in more math concepts and will practice transferring that experience to pencil and paper. Depending on the students' prior abilities, you may need to approach spatial visualization more formally with some students than with others.

When students act out a problem, they should arrive at a solution once the acting is done. As the solution unfolds, each person involved should be actively checking the problem constraints. Students will need to review the constraints after they reach a solution to verify it. Any recording should wait until after the solution is reached and verified.

©2004 Key College Publishing, *Instructor Resources: Crossing the River with Dogs,* Johnson/Herr/Kysh

Act-It-Out Day

Spend an entire class period practicing the act-it-out strategy, using the problems in Section 1: Act It Out of the student text. Divide the class into groups and assign each group a problem to do. Make sure each group has the right number of students for each problem. Every student in the group needs to take an active role in solving the problem. Give the groups 15 to 20 minutes of rehearsal. Encourage them to get up and move around the room or even go outside if that is possible at your college. Suggest (if they don't think of it themselves) that they use signs to label the actors (horse, dog, goose, coyote, and so on) and that they use something to represent inanimate objects in the problem (boat, money, figurine, and so on). After rehearsal, have the entire class sit down to be the audience and have each group come to the front of the room and act out their solution to their problem.

The following problems require the number of people indicated. You can have an extra person in a group to act as a director or narrator if you want. You probably have more problems than you need. You could use one as the Problem of the Day and skip some others.

Jackals and Coyotes: 6 people (very difficult)

Horse Trader: 3 people

Three Adults and Two Kids: 5 people (easy)

The Dog, the Goose, and the Corn: 4 people (easy)

Hoop Greeting: 10 people (*Note:* You could have a group with just a few people to rehearse, then have them get volunteers from the audience for the performance. Or don't assign this problem at first and leave it for people who finish early on other problems.)

Switching Jackals and Coyotes: 6 people

The Hotel Bill: 5 people

Persis's Gift Shop: 4 people (This problem makes an excellent Problem of the Day and might be better used as such.)

Using Manipulatives

The day after act-it-out day, put your class in their groups and have them repeat the two versions of Jackals and Coyotes, using manipulatives. Little pieces of paper marked J and C work well, as do centimeter cubes, pennies and nickels, and various other things. Then have students work in their groups on the problems in Problem Set A-2 on pages 268–72 in the student text, using manipulatives. The first problem in that set, Two Jackals Lose Their Licenses, is very difficult. Students should all have done the first two Jackals and Coyotes problems several times before they tackle this one.

Notes on Problem Set A-1

THREE ADULTS AND TWO KIDS

You may wish to have your students try a manipulatives approach to this problem. Possibly do the solution using a length of ten colored rods as the boat, a length of eight as each adult, and a length of five as each kid. This approximates one adult fitting or two kids fitting in the canoe, but not one of each. Pounds are converted to length if you work the problem out in this manner.

Notes on Text Problems—Section 2

In Section 1: Act It Out, the physical representation used people acting as animals or inanimate objects. Each person presumably moved herself or himself and had a degree of self-direction.

With models and manipulatives, most of the direction will reside in fewer hands and minds. The direction will not be in the hands of the problem objects; instead, students act as problem directors.

Students should probably wait to record their work until after they've found a complete solution. Seek the students' creativity in finding ways to record the process.

FOUR CONTIGUOUS STAMPS

This answer assumes that you start with a sheet of identical stamps. Orientation is important. Some students will have difficulty devising a systematic approach to solving this problem.

©2004 Key College Publishing, *Instructor Resources: Crossing the River with Dogs*, Johnson/Herr/Kysh

VOLLEYBALL TEAM

Problems of this type abound in any sport. They can be found in arranging batting orders; positioning soccer players; and demonstrating plays in football, soccer, and basketball.

■ **MEXICAN RESTAURANT**

Initially, the manipulatives are important because each one contains an element of the problem. Their position is also important. Their shape and orientation become more important as you tape the positionally related pieces together.

Note on Problem Set A-2

■ **PROBLEM VATS**

This problem tends to be counterintuitive for most people. Try using ten pieces of red paper as the red liquid and ten pieces of black paper as the black liquid. "Pour" from one "vat" by moving any number of the red pieces to the black. Return the same number of pieces—all red, or all black, or mixed colors—to the first "vat" so that each vat has ten pieces in it. No matter what, the number of reds in the black will equal the number of blacks in the red. Alternative manipulatives for this problem include a deck of cards (use ten black cards and ten red cards) or coins (use dimes for red and pennies for black).

Good Quiz Problems

We recommend no quiz on this chapter. It is difficult to quiz on this strategy.

©2004 Key College Publishing, *Instructor Resources: Crossing the River with Dogs*, Johnson/Herr/Kysh

Text Problems

JACKALS AND COYOTES

Three jackals and three coyotes are on a trek across the Mokalani Plateau when they come to a river filled with carnivorous fish. There is a rowboat in sight, and the party decides to use it. (Both species are known for their cleverness.) However, the boat is too small to hold any more than two of the group at a time, so they must traverse the river in successive crossings. There is one hitch, though: The jackals must not outnumber the coyotes at any time, in any place. For example, if two jackals and only one coyote are together on the western side of the river, this problem is reduced to simple subtraction: the jackals will overpower, kill, and eat the coyote. It's okay to have an equal number of each, and it's also okay to have more coyotes than jackals in a given place—neither situation poses a danger to the coyotes, and the coyotes do not pose a threat to the jackals. The trick here is to use the one small rowboat, a lot of sweat, and a little brainpower to ensure the coyotes' safety while both groups cross the river.

Close the book and find some people to act this out with. Also find some object to be physically transported across the room as the boat.

HORSE TRADER

Once upon a time, there was a horse trader. One morning, the horse trader bought a horse for $60. Just after noon, the horse trader sold that same horse back to the original owner for $70. He then bought it back again just before 5:00 p.m. for $80. By midnight he managed to sell the horse back to the original owner for $90. How much money did the horse trader make or lose on this horse?

Close the book and act out this problem before continuing. You will need something to represent the horse, some play money in different denominations, and at least two people to act out the roles of the horse trader and the original owner.

Text Problems *(continued)*

JACKALS AND COYOTES REVISITED

Solve the Jackals and Coyotes problem again, this time using manipulatives. Choose something to represent coyotes and something to represent jackals. Scraps of paper with *C* written on some and *J* on others work well. Coins, paper clips, or bottle caps also work. Determine the minimum number of river crossings. Work this problem before continuing.

©2004 Key College Publishing, *Instructor Resources: Crossing the River with Dogs*, Johnson/Herr/Kysh

FOUR CONTIGUOUS STAMPS

In how many ways can four rectangular stamps be attached together? Be sure to pay attention to the thrust of this chapter. Take care to record each configuration. Work this problem before continuing.

©2004 Key College Publishing, *Instructor Resources: Crossing the River with Dogs*, Johnson/Herr/Kysh

LETTER CUBE

Build this cube to see what letter is opposite the letter *T*. Pay attention to the orientation of the letters on the faces of the cube.

To solve this problem, draw the figure shown at right on a piece of paper (graph paper works well), cut it out, and fold it to make a cube. Then write the letters on its faces. Work this problem before continuing.

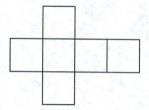

©2004 Key College Publishing, *Instructor Resources: Crossing the River with Dogs*, Johnson/Herr/Kysh

Text Problems *(continued)*

VOLLEYBALL TEAM

The volleyball team has six players: Virginia, Fannie, Helen, Dermid, Melvin, and The Prof. Using the following clues, put the players into the starting positions that will give the best rotation:

1. The players must alternate by gender.

2. Virginia is the team's best server, so she should start in the serving position.

3. Melvin and Helen are the team's setters. They must be opposite each other at all times.

4. Dermid and Fannie communicate well—it helps to put them next to each other.

5. The Prof is an effective server. He needs to be positioned so he will rotate into the serving position quickly.

(As this diagram shows, half of a volleyball court has six players; the opposing team plays on the other half of the court. From the net, there are three players in the front row and three in the back row. The server is in the right back corner. The players rotate in a clockwise manner. Players are considered to be opposite if they are three positions apart. So server 1 is opposite server 4, server 2 is opposite server 5, and server 3 is opposite server 6.)

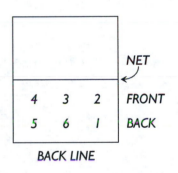

Work this problem before continuing.

©2004 Key College Publishing, *Instructor Resources: Crossing the River with Dogs,* Johnson/Herr/Kysh

©2004 Key College Publishing, *Instructor Resources: Crossing the River with Dogs,* Johnson/Herr/Kysh

Text Problems (continued)

NUMBER PUZZLE

Use the digits 0, 1, 2, 3, 4, 5, 6, 7, 8, and 9 once each to fill in the blanks of this puzzle:

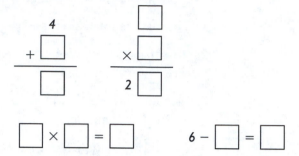

$$\square \times \square = \square \qquad 6 - \square = \square$$

Work this problem by cutting up little pieces of paper and labeling them with the numbers 0 through 9. Then manipulate them by moving them around in the puzzle until you find an arrangement that works. There is more than one possible solution.

MEXICAN RESTAURANT

Four friends (one is named Janie) went out to dinner at a Mexican restaurant. The hostess seated them in a booth. Each ordered a different meat (pork, mahimahi, beef, or chicken), and each ordered a different kind of Mexican dish (tostada, burrito, fajita, or chimichanga). Use the clues that follow to determine what dish each person ordered, the kind of meat it contained, and where each person was sitting.

1. The person who ordered mahimahi sat next to Ted and across from the person who ordered a burrito.

2. Ken sat diagonally across from the person who ate the fajita and across from the person who ordered beef.

3. The person who ordered a chimichanga sat across from the person who ordered chicken and next to Allyson.

Work this problem before continuing.

Notes on Optional Section: Conversions

The Conversions section that follows on page 185 is intended to serve as a problem-solving bridge back to the real world. This section could be particularly valuable for a class of future teachers.

- It takes abstract problems and gives them physical representations.
- It gives some tips on how to transform certain types of problems into problems that are easily solved with manipulatives.

Unless a problem explicitly suggests manipulatives, many people do not immediately think of that strategy for solving it. This section is intended to show the connection as well as teach how to give routine problems a manipulatives slant.

Several of the problems require colored rods, also known as Cuisenaire rods. If you don't have access to these rods, you can use the template on page 184 to make your own. The shortest rods are white and are 1 cm long; the longest, orange, are 10 cm long.

Pages 185 to 190 are set up as blackline masters for student pages. Here are the answers to Problem Set: Conversions.

9. Trains from Salt Lake City: 4 hours 36 minutes after the first train left

10. Catching Remzi: 2:00 p.m.

11. Inchworm: 3 minutes

12. Can Do, 9 six-packs and 3 eight-packs

13. Soda: Answers will vary; there are 27 possible combinations.

Orange		
W	Blue	
Red	Brown	
Lt. Green	Black	
Purple	Dark Green	
Yellow	Yellow	
Dark Green	Purple	
Black	Lt. Green	
Brown	Red	
Blue	W	
Orange		

Conversions

Work these six problems.

1. HOLIDAY BLOCKS

Use red and light-green colored rods. What are the possible totals of each color required to exactly fill this space?

2. PUNK BLOCKS

To do this problem, you are allowed to use exactly 13 colored rods. Some can be red; the rest must be purple. How many of each color are needed to exactly fill this space?

3. SPRING BLOCKS

Use only purple and yellow colored rods. Make a string of purples that is the exact same length as a string of yellows, using one more purple than yellow.

4. TRIKES AND BIKES

Mr. Wheeler has tricycles and bicycles in his garage. The total number of wheels on those vehicles is exactly 27. What are the possible combinations of trikes and bikes?

5. MORE DUCKS AND COWS

Janice counted a total of 13 heads and 36 legs in a barnyard. She couldn't remember how many of the animals were ducks and how many were cows. How many of each were there?

6. NEXT TRAIN EAST

A train leaves Roseville heading east at 40 miles per hour at 6:00 a.m. Another leaves on a parallel track at 50 miles per hour at 7:00 a.m. At what time will the trains be the same distance from Roseville?

Conversions *(continued)*

The problems on the previous page are pairs of problems that are set in different situations (and worded differently). Problems 1 and 4 are a pair, 2 and 5 are a pair, and 3 and 6 are a pair. The two problems in a pair share the same mathematics but are worded to suggest different methods for solving them.

The purpose for pairing problems in this way is fourfold: (1) to show that problems can be worked in different ways, (2) to show that the wording of a problem will suggest certain solution strategies, (3) to highlight the relationship between abstract mathematics and the physical world, and (4) to introduce methods for converting problems into a format and wording that suggest using manipulatives to solve them.

Here's an analysis of the problems: Problem 1 asks you to figure the number of red rods (two squares) and light-green rods (three squares) that would fill a space covering 27 squares. Problem 4 asks you to use bicycles (two wheels) and tricycles (three wheels) to get a total of 27 wheels.

In Problems 2 and 5, you need to come up with a combination of thirteen 2's and 4's that gives a total of 36. Problem 2 asks you to fit some blocks into a space (the area of that rectangle is 36 square centimeters, and the blocks are either 2 or 4 square centimeters long). In Problem 5, the 2's and 4's represent the number of legs on animals.

Finally, Problem 6 asks you to work a train problem that involves the formula *distance = rate × time*. But in Problem 3 you use manipulatives that could represent the distances per hour in Problem 6, and you use one more purple rod (4 cm long to represent 40 mi/hr) than yellow rod (5 cm long to represent 50 mi/hr), which could represent a train traveling one hour longer.

When problems can be worked in various ways, your choice of method partly depends on how the problem is presented. When the problems on the previous page were stated in terms of manipulatives, chances are you worked the problems using the manipulatives approach. When the problems were presented without mention of manipulatives, you probably worked them in some other way.

A number of problems can be rewritten to instill a suggestion for (or against) being solved with manipulatives. It is no accident that math problems can be solved with manipulatives. Most math is based on real-world models. Therefore, changing problems back into real-world interpretations really isn't that big a jump. See pages 264–267 of the student book for suggestions about when to use manipulatives.

Convert the next two problems into problems that are slanted toward using manipulatives.

Conversions (continued)

Rewrite these problems in a way that suggests using manipulatives as a solution strategy.

7. MALCOLM'S WHEELS

Malcolm used to brag that he had 11 vehicles at his estate; all were either motorcycles or cars. In all, his vehicles had 36 wheels. How many of each did he have?

8. A DRIVE TO THE LAKE

My grandfather told me about the time he, his sister, and his cousins drove to the lake from his cousins' house. They took separate cars. My grandfather and his sister drove 30 miles per hour. His cousins left two hours later and drove 40 miles per hour. Both cars arrived at the lake at about the same time. What was the approximate distance from his cousins' house to the lake?

Conversions *(continued)*

There are many correct ways to rewrite each of these problems to slant toward a manipulatives solution.

For Malcolm's Wheels, the critical elements are that there are 11 items, each of which has a value of 2 or 4 (2 wheels or 4 wheels). The total value is 36. The problem could easily be changed in the following ways:

1. The wheels become centimeters, and the total is a length of 36 centimeters. The total is made up of groups of 2 and 4. Therefore, string together 2-rods (red) and 4-rods (purple) in the length that totals 36 centimeters. Use exactly 11 rods.

2. The wheels become square centimeters, and the total is an area of 36 cm². The total is made up of groups of 2 cm² and 4 cm². (You can also consider these areas in terms of rectangles, with a uniform width of 1 cm.) Therefore, fill in a rectangle whose area measures 36 cm², using exactly 11 smaller rectangles, each measuring either 2 cm² or 4 cm².

3. Use dimes and nickels. The nickels count as motorcycles, and the dimes are considered cars. Since the total value is 36, which could be made up of 18 motorcycles, consider the total value to be 90 cents (18 motorcycle-nickels × 5 cents each). Therefore, using exactly 11 coins, made up of nickels and dimes, make 90 cents worth of change.

Note that this problem *could* be acted out. Invite a number of friends over, making sure to include a combination of cyclists and motorists. (Anyone under the legal driving age may consider inviting friends with scooters and wagons.) Park the vehicles in 11 parking spaces, and guess and check until you have exactly 36 wheels in the 11 spaces.

For A Drive to the Lake, the critical elements are the average speeds (30 and 40 miles per hour) that both cars use to go the same distance and that one car drives 2 hours longer than the other.

1. A possible rewrite involves using 3-rods (light green) and 4-rods (purple), which represent rate in 10-miles-per-hour segments. There need to be two more 3-rods than 4-rods. Make strings of rods. The 3-rod string needs to be the same length as the 4-rod string to represent that both cars drove the same distance.

Conversions *(continued)*

2. Here is another way to write this problem for manipulatives: How many pennies would it take to balance a scale if (a) you can put pennies on one side only in groups of 30 and on the other side only in groups of 40, and (b) you have to use exactly two more groups of 30 pennies than you use of 40 pennies?

In the second approach, rather than balancing the groups of pennies on a scale, you could lay them out in lines to see how many it would take to make them the same length. This approach would work even if the problem were written like this:

> My grandfather told me about the time he, his sister, and his cousins drove to the lake from his cousins' house. They took separate cars. My grandfather and his sister drove 28 miles per hour. His cousins left one and one-half hours later and drove 35 miles per hour. Both cars arrived at the lake at about the same time. What was the approximate distance from his cousins' house to the lake?

The concepts for writing this as a manipulatives problem are the same: Use groups of 28 and 35 pennies. There must be one and a half more groups of the 28 pennies than groups of the 35 pennies. The two strings of pennies must be the same length. At this point, however, using manipulatives may not be practical. In order to solve this problem with pennies, you need a big pile of pennies.

©2004 Key College Publishing, *Instructor Resources: Crossing the River with Dogs*, Johnson/Herr/Kysh

Problem Set: Conversions

Make specific suggestions for modeling these problems using manipulatives. Identify the critical elements of each problem and how each would be represented in your model. Then solve the problem using your manipulatives.

9. TRAINS FROM SALT LAKE CITY

Two trains left Salt Lake City, one heading east and the other heading west. The first train is traveling at 40 miles per hour. The second train left one hour later and is traveling at 60 miles per hour. When will the two trains be 400 miles apart?

10. CATCHING REMZI

At 8:00 a.m., Remzi left Chicago for a convention, driving west on Interstate 80. He usually averages about 50 miles per hour when he drives on freeways. Natasha and Jessica, who are going to the same convention, left an hour later but were going to try to average 60 miles per hour. What time do they catch up to Remzi?

11. INCHWORM

An inchworm that is 6 inches long travels at the rate of 6 inches per minute. How long does it take the inchworm to crawl completely across a piece of paper that is 12 inches long?

12. CAN DO

Kathy, Nate, and Adam are buying some Meadow Dew soda at Mega-Save Drug Store. Right now there is a special promotion sponsored by the bottler. Some of the six-packs have six sodas, but some special promotional six-packs are actually eight-packs. Kathy, Nate, and Adam are buying a total of 12 packs of six-packs and eight-packs that give them a total of 78 cans. How many of each type of pack are they buying?

13. SODA

Marisan reported back to her friends that she bought 288 ounces of soda. Soda comes in cans (12 oz), bottles (16 oz), and quart-bottles (32 oz). What are the possible combinations she bought?

TEACHING RESOURCES

©2004 Key College Publishing, *Instructor Resources: Crossing the River with Dogs*, Johnson/Herr/Kysh

Problem Set A, Version 2

Solve each problem by acting it out or using manipulatives.

1. HANUM'S ISLAND

Hanum and her friends went to play on an island about ¼ mile offshore at the lake. They have a sailboat, but the sailboat is very small and can take only a maximum of 220 pounds at a time. Hanum weighs 120 pounds, Shelby weighs 100 pounds, Elke weighs 95 pounds, Carmen weighs 110 pounds, and Lester weighs 140 pounds. How many trips will it take to get Hanum and her friends off the island? (Consider a trip to be one way between the island and the mainland.)

2. ANOTHER MAGIC TRIANGLE

To do this magic triangle, you must use the digits 1 through 6, once each. Tear up little pieces of paper and write the digits 1 through 6 on them. Place one digit in each square to give each side of this triangle a sum of 12.

3. ROOKIE PURCHASE

Max paid only $1,500 for a Willie Mays rookie card worth $1,800. He later sold it for $2,000, but the buyer came back and showed him that it was actually a fake. Max gave the customer back his money. Who lost how much on this?

4. LAWN CHAIR AND BARBECUE

Sammie bought a lawn chair and a barbecue grill at his neighbor's garage sale for $20. He then sold the lawn chair to Wilma for $9. He sold the barbecue grill to Marc for $12. Marc later sold the grill to Wilma for $14, who then sold both items back to Sammie for $22. How much money did each person make or lose on the grill and the lawn chair?

©2004 Key College Publishing, *Instructor Resources: Crossing the River with Dogs,* Johnson/Herr/Kysh

Problem Set A, Version 2 *(continued)*

5. STAMPS

How many different ways are there to make 29 cents in stamps if the stamps all cost 8, 5, or 2 cents each? Use colored rods to solve this problem.

6. LETTER CUBE

Build this cube to see what letter belongs on the blank face. Pay attention to the orientation of the letters.

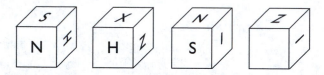

Draw a figure like the one below on graph paper and cut it out to assist with this problem.

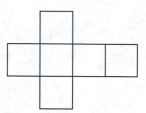

7. SIX AND FOUR

What is the minimum number of 90-degree rotations needed to bring this standard pair of dice to the point where each die has the six spots in front and the four spots on top?

TEACHING RESOURCES

©2004 Key College Publishing, *Instructor Resources: Crossing the River with Dogs,* Johnson/Herr/Kysh

Problem Set A, Version 2 *(continued)*

8. A HEART IN THE RIGHT PLACE

There is a heart to the right of a diamond. There is a diamond to the left of a diamond. There is a heart to the left of a diamond. The sum of the card values is 11. One of the card values alone is greater than the sum of the other two card values. The cards are arranged from least to greatest in order from left to right. There are no aces. What are the cards?

9. TWO DICE

Two dice are placed side by side. From the front, the dice show a sum of nine. On top, the dice show a sum of four. The sum of the right and left ends is six. How are the dice positioned? Sketch your answer.

10. JUST A PEW KIDS

Dad and four kids are sitting in a church pew. Brenna wants to sit away from Shawn because he squirms so much during the service. Shawn doesn't want to sit by his sister Tara. The youngest, Krista, needs to sit by Dad and Dad likes to sit by the aisle so that he can take Krista out when she gets antsy without making a disturbance. How should they sit in five consecutive seats in church?

11. FIVE AT THE TABLE

Five people are sitting at a cafeteria table that has room for five people on each side. Each person is either sitting immediately next to or directly across from another person, or both. How many seating arrangements are possible?

12. ANOTHER MAGIC SQUARE

The sum of rows in a magic square is 15. The sum of each diagonal is 15. The sum of each column is also 15. Use the digits 0, 2, 3, 4, 5, 6, 7, 8, and 10 and find the proper location of each. (There is more than one correct solution.)

Problem Set A, Version 2 *(continued)*

13. **CARDS IN TWO ROWS**

Three cards were placed faceup in a row. Three more cards were placed on top of them, also faceup, but you could still see the bottom-row cards. There was a king on an ace. The queens were not in the same row. The jack was to the left of an ace. One of the aces was to the right of a queen, and the other was to the left of a queen. The jack was under one queen and to the left of another. The king was to the left of an ace. No two cards of the same rank were in the same stack. Any cards to the left or right of one another have to be in the same row. How were the cards arranged?

14. **BLUEGRASS AT DINNER**

At a musicians' conference, several friends got together for some dinner before the evening performance. As fate would have it, three different bluegrass groups were represented. Each group consisted of a guitarist, a banjo player, and a fiddler (violinist). Determine who is in what group. (By the way, fiddles have four strings, banjos have five, and these guitars have six.)

1. Irving and Ben play banjo.

2. Kathleen and Sue are in the same group.

3. George and Ben haven't seen each other since the last conference.

4. Kris and Irving are in the same group.

5. Emily's instrument has fewer strings than both Ben's and Kris's. Kathleen's instrument has more than four strings, and Sue's instrument has fewer than six.

6. Kris and Lucille play the same instrument.

7. Marty, who plays banjo, always feels inspired listening to Lucille's group.

Who plays the fiddle in Marty's group? Who plays banjo in Emily's group? What instrument does George play, and who else is in his group?

TEACHING RESOURCES

©2004 Key College Publishing, *Instructor Resources: Crossing the River with Dogs,* Johnson/Herr/Kysh

15. **GOOD NEWS FOR CUBES**

Suppose you had two figures, each comprising three squares as shown below.

In Figure 1 below, *A* and *Y* were placed next to each other. The figure does not form a cube when folded.

In Figure 2, *A* and *X* were placed next to each other. The figure does form a cube when folded.

Figure 1

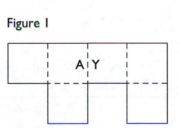

Figure 2

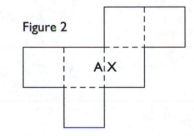

1. Can a cube be folded by attaching *A* to *Z*?

2. By attaching *B* to *Y*?

3. By attaching *C* to *X*?

Problem Set B, Version 2

1. DUELING PINS

Three married couples decided to have a bowling competition. The first game featured the Sanchezes against the Wojics. The second game matched up the Wojics and the Carsons. The Carsons bowled against the Sanchezes in the third game. It was hard to tell who was pushing the competition more, the women (Lani, Fawnda, and Rhonda) or the men (Derrek, Hank, and Bill). The winner was declared on the basis of the total score for the two games. From these clues, determine the full names of the couples and the scores of each game. Clues 1 to 5 refer to single games. Clue 6 refers to the whole match.

1. Hank and his wife lost to Fawnda and her husband.

2. Lani and her husband beat the Sanchezes.

3. The Carsons won one game, 258 to 217.

4. Rhonda and her husband lost to the Wojics, 243 to 290.

5. Bill and his wife lost one game, 211 to 281.

6. Lani and her husband came in second in the competition with a total of 507 pins. The winning couple had 524 pins.

2. STOPLIGHTS

On San Francisco's 19th Avenue, there is a stoplight on every corner. The stoplights are all timed, and they stay green for 1 minute, yellow for 5 seconds, and red for 1 minute. The intersections are 200 yards apart. The lights are timed so that each light turns green 10 seconds after the previous one. You are stopped at a stoplight. When the light turns green, it takes you 7 seconds and 50 yards to reach a speed of 30 miles per hour. From then on, you travel a constant 30 miles per hour. How many stoplights can you drive through before you have to stop for a red light? Assume you will go through a yellow light if it has been yellow for only 2 seconds when you reach it, but you will stop if it has been yellow for more than 2 seconds when you reach it.

©2004 Key College Publishing, *Instructor Resources: Crossing the River with Dogs*, Johnson/Herr/Kysh

Problem Set B, Version 2 *(continued)*

3. TICKET PRICES

Four theaters are on the same street in Miami. One day the four theater owners were discussing their ticket prices. The movie tickets were cheapest, followed by the community theater tickets, the dinner theater tickets, and the symphony tickets. They noticed that if three of the ticket prices were averaged and then added to the ticket price of the fourth theater, the numbers 25, 27, 31, and 39 would result, depending on which price you left out of the average. What are the ticket prices?

4. CAPICUA NUMBERS

Your Spanish cousin is in town for a week. She tells you about "capicua" numbers: numbers that read the same forwards or backwards. (In English they are called "palindromes.") For example, 55 and 94549 are capicua numbers. It is considered good luck to get one for your ticket number on the Madrid subway. How many capicua numbers are there between 1 and 100,000?

5. A FAST (?) SPACESHIP

Light travels at a rate of approximately 186,000 miles per second. The North Star is approximately 460 light-years away from Earth. (A light-year is the distance light travels in one year.) How many years would it take a spaceship traveling at 1 billion miles per hour to reach the North Star?

©2004 Key College Publishing, *Instructor Resources: Crossing the River with Dogs*, Johnson/Herr/Kysh

Problem Set B, Version 3

1. ODD-ODD NUMBERS

Odd-odd numbers are numbers that not only end in an odd digit: All of their digits are odd. For example, 3,515; 73; 1; and 97,757,931 are all odd-odd numbers, while 3,576; 64,428; and 997,553,121 are not. How many odd-odd numbers are there between 1 and 1,000,000?

2. THE BRIDGE TOURNAMENT

Three married couples named Friedrich, Muir, and Hatfield decided to have an informal three-way bridge tournament. Each couple played each of the other couples once in a series of four-game matches. The third couple would watch TV while the other two couples were playing bridge. It turned out that each couple won one match and lost one, and so the winner was declared to be the couple who won the most points altogether. The women's names were Anne, Mary, and Jean. Their husbands were Randy, Jim, and Dave. Using the clues below, determine the full names of the couples and figure out their order of finish in the tournament. Each clue refers to a single match.

1. The Hatfields lost to Anne and her husband, 320 to 410.

2. Jean and her husband beat the Muirs.

3. Randy and his wife lost, 80 to 280.

4. Dave and his wife beat the Friedrichs.

5. Jim and his wife lost to Mary and her husband, 400 to 560.

3. DISCOUNT TICKETS

Kristi and her adult friends made plans to go to the amusement park. She looked in the paper and saw that at the regular ticket price it was going to cost her $204 to buy enough tickets for herself and all her friends. She started calling her friends and found out that none of them could go. So she changed her plans to a family-oriented day. She planned for half as many adults as she had earlier planned and a bunch of her teenage and preteen cousins. With all these adults, teens, and children, she needed 15 more tickets than she had originally planned for her adult friends. Surprisingly, it was still going to cost her $204 at regular prices. The next day she went to the park to buy the tickets. It turned out that the park was having a sale, with all tickets 20% off, so she actually spent less than $204. There were three kinds of tickets: adult, teen, and child. For the price of an adult ticket, she was able to buy a teen ticket and a child ticket. Teen tickets were $4.00 on sale, and she bought 5 of them. How many children's tickets did Kristi buy?

©2004 Key College Publishing, *Instructor Resources: Crossing the River with Dogs*, Johnson/Herr/Kysh

4. DRIVING ON THE FREEWAY

In my driver's education class last week, my teacher was talking about driving on the freeway. He said that you should stay one car length behind the car in front of you for every 10 miles per hour that you drive. (So, if you were driving 60 miles per hour, you should be six car lengths behind.) I went home and told my dad, and he said that his driver's education teacher had told him that he should stay 2 seconds behind the car in front of him. Assuming that a car length is 20 feet and you are traveling 60 miles per hour, how many car lengths behind should you stay, using Dad's 2-second rule?

5. CAKE RECIPE

A recipe for four-egg cake calls for the ingredients listed below and is supposed to be baked in three 9-inch-diameter round cake pans. Unfortunately, you have only three 10-inch-diameter round cake pans. You want to make your cake about the same height as it would have been in a 9-inch cake pan. Determine the amounts you need for each ingredient. *Hint:* Do the eggs first.

Ingredients

2⅔ cups cake flour

2¼ teaspoons baking powder

1 cup butter

2 cups sugar

4 eggs

1 cup milk

Problem Set B, Version 4

1. **BATTING AVERAGE**

I went on vacation recently. When I left, my favorite baseball player had 138 hits in 423 at bats for a batting average of .326. He had played in 110 games so far. I was gone for 12 days, and the team played a game every day. When I came back, he was batting .309. What was his probable batting average for the games that I missed? (Assume he played in every game.)

2. **PYRAMIDS**

The eight cheerleaders from Stackem High liked to do various acrobatic stunts in which they would climb on each other's shoulders to form pyramids. However, to prevent injury they had agreed that no person would stand on the shoulders of someone who weighed less than herself. Three of their favorite pyramid schemes are shown below. *Note:* The far right shows vertical pyramids of one person on another's back. Scheme 1 shows Kelly on Barbara's back. Scheme 3 shows Debbie on Barbara's back, with Barbara on Carmella's back. Use Schemes 1, 2, and 3 to rank the girls in order from lightest to heaviest.

Scheme 1

Lori

Carmella   Diane   Kelly

Kate   Lisa   Debbie   Barbara

Scheme 2

Diane   Lori   Lisa

Carmella   Barbara   Debbie   Kelly   Kate

Scheme 3

Debbie

Lori   Kelly   Barbara

Lisa   Kate   Diane   Carmella

Problem Set B, Version 4 *(continued)*

3. STEP ON A CRACK

Sohrab is taking a long walk. He always finds it interesting to see how many sidewalk cracks he stepped on. The cracks are 5¼ feet apart. The cracks are very thin. His stride is 2¾ feet long (measured from heel to heel or toe to toe). His feet are 12 inches long. Assume his foot has to overlap a crack by at least an inch for him to count it. If he starts his walk with both heels just barely past a crack, what fraction of the cracks will he step on during his walk? (Consider the beginning of a cycle to be when his heel is just past a crack. End the cycle when his heel is just past a crack again. The cycle is the number of cracks—don't count both the beginning and ending cracks, just count one of them. The fraction of cracks he steps on is the number of cracks he steps on divided by the number of cracks in the cycle.)

4. START YOUR ENGINES

Thursday night is classic car night at Jasper's Hamburgers. Every night about 150 classic cars are on display, there's oldies music, and people eat hamburgers and french fries. It's a lot of fun. The restaurateurs decided to sponsor several vehicles at a regional car show but had trouble deciding which cars to choose. So Marsha (the restaurant owner) devised a plan: She told everyone to turn on their headlights. Marsha then counted every other car (the second car, the fourth, the sixth, and so on). The owners of those cars had to turn off their headlights. Marsha then counted every third car (the third, sixth, ninth, and so on) and told the car owners who had their headlights on to turn them off. Those who had their headlights off had to turn them on. She continued in this fashion, counting by each subsequent number (next by 4's, then by 5's, then 6's) until she had finished counting by 153's. (There were 153 cars, so counting by 153's didn't take very long.) Each time she counted off cars, the cars she pointed to switched from lights on to lights off, or vice versa. Which cars still had their headlights on when Marsha was done? Explain why those particular lights were on.

5. **RODGERS AND HAMMERSTEIN**

At our house, we love movie musicals written by Richard Rodgers and Oscar Hammerstein. We own five of their most famous movie soundtracks on compact disc (CD): *Oklahoma* (1955), *Carousel* (early 1956), *The King and I* (late 1956), *South Pacific* (1958), and *The Sound of Music* (1964). I like to listen to them all in one afternoon and to listen to them in chronological order. Fortunately, I own a five-CD carousel that will play discs 1 through 5 continuously. Unfortunately, the discs were put in the machine by my son, who didn't understand my preference for playing them in chronological order. The current order is disc 1, *South Pacific*; disc 2, *The King and I*; disc 3, *The Sound of Music*; disc 4, *Oklahoma*; disc 5, *Carousel*. I need them to start with *Oklahoma* in the disc 1 spot all the way through *The Sound of Music* in the disc 5 spot. To switch the discs, I start at disc 1, take that disc (which is *South Pacific*) out, press "Disc Skip" whatever number of times is necessary to move to the disc I want to exchange, then exchange that disc for the one in my hand, and so on. (I can exchange only one pair of discs at a time.) The carousel rotates from disc 1 to discs 2, 3, 4, and 5 and then returns to disc 1. How can I get the discs in the right order and press "Disc Skip" the least number of times? The CD player needs to finish on disc 1, so that I can start playing them with *Oklahoma*.

▮▮ Work Backwards

Working backwards is analogous to the process used in solving algebraic equations. If you think of an algebraic expression as a variable with a bunch of operations performed on it, you can think of solving an equation as working backwards to isolate the variable you started with. Working backwards is a commonly used strategy in real-life problem solving, too. Budgets are often made by working backwards from how much money is available. Scheduling is often performed by starting with the question "When does it need to be done?" Then, by working backwards through the project, you can determine when different tasks need to be started. An accident investigator must work back from the end result to see what transpired.

Working backwards is a way of Changing Focus. Whereas most problem solutions describe a result of a number of actions or operations, the solution to a problem solved by working backwards usually describes a state that existed *before* the action of the problem occurred. You must start by focusing on the end state of the problem and work backwards from there.

Students can write their own problems that can be solved by working backwards. Cookies are a good problem subject. Suggest that students choose a number of cookies and then describe what happens until they reach some number of cookies left. Students can trade problems to practice solving them.

When teaching this strategy, you will need to emphasize two things: doing the opposite operation and doing these operations in reverse order. Students will likely find doing operations on fractions in reverse the most difficult concept. For example, ⅕ of 100 is 20, so if someone ate ⅕ of the jelly beans, there would be only 80 left. However, you need to recognize that the 80 is ⅘ of the original number. To get back to 100, you need to multiply by ⁵⁄₄ or to add ¼ (not ⅕) of 80. Diagrams and pictures of fractional pieces will help with this strategy. The following page presents an example of using diagrams to work the fractions backwards.

"Barry ate ⅗ of Willie's cookies. Willie had 8 left."
Draw a diagram with five parts in it (to show the "before" state). The 8 remaining cookies are the ⅖ left, so they are split into two of the boxes.

Fill in the other boxes with the same number. Each fifth represents 4 cookies; hence all 5 fifths make 20 cookies.

Point out to students that solutions found by working backwards can be checked just like solutions to equations: Start with your answer at the beginning of the problem, perform all the actions working forwards, and see if you arrive at the expected result.

For a good article about this type of problem, see "A Puzzle for Pirates" by Ian Stewart from the Mathematical Recreations column of the May 1999 issue of *Scientific American*. The article is available at Ian Stewart's personal Web page as well (http://members.aol.com/istewjoat/sampleMR.html#sampleMR).

Good Quiz Problems

For a one-problem quiz on this strategy, consider one of these problems from Problem Set A, Version 2:

  2. Orange You Hungry?

  5. Brown Thumb

  9. Political Buttons

  10. Fruit Stand

Text Problems

POOR CHOICES

The night before their debut in Carnegie Hall, the dancers stayed at a hotel on 57th Street. Fourteen members of the ballet company went to an all-night card room to play poker. Half of the remaining dancers went to Madison Square Garden for a special midnight professional wrestling show featuring Buff Bargle. After about an hour, 6 of the dancers who had gone to play poker came back to the hotel broke. The 11 dancers now at the hotel went to bed and got enough sleep, but the rest of the dancers were tired for their debut the next day. How many dancers were in the ballet company? Work this problem before continuing.

DAD'S WALLET

Dad went to the ATM on Wednesday of spring break and withdrew some money. On Thursday morning my brother borrowed half of Dad's money to open a checking account, because he was always short of money. On Friday I needed some money for a date, so I borrowed half of what remained. My sister came along next and borrowed half of the remaining money. Dad then went to gas up the car and used half of the rest of his money, and he wondered why he had only $15 left. How much money did he start with in his wallet? Don't forget to reverse the actions and the order. Don't read on until you've worked this problem.

NUMBER TRICK

Start with a number between 1 and 10.
    Multiply the number by 4.
    Add 6 to the number you have now.
    Divide by 2.
    Subract 5.
    Tell me the number you end with, and I'll tell you the number you started with.
    Two students, Glenda and Sonia, played this game. Sonia started with a number, did the arithmetic, and told Glenda that she had ended with 12. Glenda then figured out what number Sonia started with. What number did Sonia start with? Work this problem before continuing.

Text Problems *(continued)*

MINTS

After dinner, three friends at a restaurant paid their bill and noticed a bowl of mints on the front counter. Sean took one-third of the mints but returned four because he had a momentary pang of guilt. Faizah then took one-fourth of what was left but returned three for similar reasons. Eugene then took half of the remainder but threw two green ones back into the bowl. The bowl had only 17 mints left when the raid was over. How many mints were originally in the bowl? Work this problem before continuing.

Problem Set A, Version 2

I. MARY'S CAR LOT

Mary decided to focus on selling luxury cars and needed to sell off one line of sports cars: the LaMancinis. She held a clearance sale. The first week of the sale, she and her associates sold half the LaMancinis. The second week, they again sold half the remaining sports cars on the lot and one more. Mary and the other three sales representatives then each chose one of the remaining sports cars to keep, and there were none left over. How many LaMancini sports cars were on the lot at the start of the sale?

2. ORANGE YOU HUNGRY?

Leona set up a fruit stand to sell some of her crop of oranges. The first person who came by bought ⅓ of her oranges. The second person bought 4. The third person then bought ¼ of the remaining oranges. Leona took the last 15 oranges home and made orange juice. How many oranges did Leona have when she set up the fruit stand?

3. TURNPIKE

Maury was driving down the turnpike. Every time he got to a tollbooth, he had to pay $1. He went through one tollbooth and then stopped at a restaurant. He spent half of his money on dinner and then continued on the highway. A while later, he went through another tollbooth and then stopped to buy some snacks for $2. He then passed through another tollbooth, spent half of his money on a soda, and then spent his last dollar on a tollbooth, just before turning off at his exit. How much money did Maury have when he started on the highway?

4. THE MONOPOLY GAME

In the last two times around the game board, Wendell thought he was a goner. First, he had to pay ½ of his money in rent to somebody. He then had to pay $25 because of a "Chance" card. After that, he landed on another property where he had to pay ⅗ of his remaining money in rent. Fortunately, on his next turn he passed "Go" and collected $200. However, he landed on a property that whacked him for ½ of his money. Next he got a "Chance" card and landed in jail. After he paid $50 to get out, the poor guy had only $70 cash left—no property, no nothing. How much did he have at the beginning of this narrative?

©2004 Key College Publishing, *Instructor Resources: Crossing the River with Dogs*, Johnson/Herr/Kysh

Problem Set A, Version 2 *(continued)*

5. BROWN THUMB

I can't believe how bad my brother, Reuben, is at growing things. He bought a whole bunch of plants at the nursery the other day. Right away, five died. Then our dog dug up $2/9$ of them. Out of the remaining plants, $1/2$ of them died soon afterward. Then a rabbit came and ate $1/2$ of what was left and $1/2$ a plant more. Reuben had only three plants left then. How many did he buy at the nursery?

6. JELLY BEAN RABBITS

My sister, Lilly, and I used to have a great time playing with jelly beans. First we would place them in groups on the living room floor. Then we would pretend they were rabbits and we were hawks. We would swoop down and eat them up. One time in particular, Lilly got the first shot: She ate $1/5$ of the jelly beans. I went next, and I ate $1/6$ of the ones on the floor. Lilly then let out a shriek and ate $2/5$ of the remaining beans. I then ate $2/3$ of the ones left. After Lilly ate $1/2$ of the remaining jelly beans, there were two yellow ones left, which we shared. How many jelly beans did each of us eat?

7. WINNING GOLDFISH

Triva went to the arcade at the state fair to win some goldfish. She already had some at home, but she wanted more. Right away, Triva won enough goldfish to double her stock. However, her mom made her give four to her cousin. She put her new ones in the fish tank with the others, but by the next morning, $1/2$ of her goldfish had died. Triva's friend Keisha gave her six more. Unfortunately, the next morning $2/3$ of her goldfish had died. Triva was left with two goldfish after having given one to a neighbor. How many goldfish did Triva start with?

8. PRETTY FISHY

Triva took the fishbowls she'd bought to the flea market. In the first hour, she sold $1/3$ of them and $1/3$ of one more. In the second hour, she sold $1/2$ of them and $1/2$ of one more. In her third hour there, she sold $1/3$ of them and $1/3$ of one more. The next hour, she sold $1/2$ of them and $1/2$ of one more. Finally, she sold the last two and went home. How many fishbowls did Triva sell?

©2004 Key College Publishing, *Instructor Resources: Crossing the River with Dogs,* Johnson/Herr/Kysh

Problem Set A, Version 2 *(continued)*

9. POLITICAL BUTTONS

Gil buys and sells political buttons on the Internet. He really likes Barry Goldwater buttons. (For those who don't know, Barry Goldwater ran for president in 1964.) A while back, Gil purchased a whole bunch of Goldwater buttons. Some time later, he sold ⅗ of the buttons. The next day, he sold 6 buttons. The following day, he sold ¼ of the remaining buttons. A few hours later, he sold 8 more buttons. The next week, he sold ½ of all the remaining buttons and ½ a button more. After all that, he had 3 buttons left. How many buttons did Gil purchase originally?

10. FRUIT STAND

Ashley set up a fruit stand in order to sell some of her crop of watermelons. The first person who came by bought ⅓ of her watermelons. The second person bought 6 watermelons. The third person then bought ¼ of the remaining melons. The last person bought ½ of all the remaining melons and ½ a melon more. Ashley took the last 7 watermelons home. How many watermelons did Ashley have when she set up the fruit stand?

©2004 Key College Publishing, *Instructor Resources: Crossing the River with Dogs*, Johnson/Herr/Kysh

Problem Set B, Version 2

1. CHICKEN EGGS

If a chicken and a half lays an egg and a half in a day and a half, how many eggs will 6 chickens lay in 12 days?

2. HEARTBEATS

Generally speaking, the fewer times your heart beats every minute, the healthier you are, and a longer life span will result. Joanne and Ken compared their heart rates. Joanne had been exercising every day for at least a year. Her heart rate during the 45 minutes each day that she exercised was 150 beats per minute. But her heart rate dropped to 60 beats per minute for the rest of the day. Ken didn't exercise at all and had a daylong average heart rate of 72 beats per minute. During the last non-leap year, whose heart beat more times and how many more times did it beat?

3. PENNIES IN THE SAND

I hate taking my little brother, Cal, to the park. All he does is find ways to cause trouble. For example, the other day he took a bunch of pennies to the park. He promptly lost ⅖ of them and cried. I stopped swinging and helped him look for them. He managed to let three more fall out of his pocket while we were looking. When I tried to grab them before they sank in the sand, he threw a temper tantrum and lost another ⅓ of the pennies he had left. I gave him ten pennies from my pocket to try to stop him from crying, but that didn't help. He then got mad and started throwing them. He threw away ¾ of his remaining pennies and ended up with seven. We looked for them some more but found a lot of cigarette butts and no pennies. How much money did my brother lose at the park? I don't know, but I would have gladly paid him that much not to go to the park. Find out how much Cal lost at the park.

4. JOGGING

I was walking down the street one day, and a jogger passed me. I was curious to know whether I could figure out how fast the jogger was running, so I started paying attention to where the jogger was. When she passed a light pole a little ways in front of me, I began to count seconds. It took me 9 seconds to get to the pole. When I reached the pole, I looked up to see that she was just passing a fire hydrant. I again counted seconds until I reached the fire hydrant, and it took me 11 seconds. When I reached the fire hydrant, she was passing a parking meter. It took me 14 seconds to reach the parking meter. I knew that I walked two paces in 1 second, and my pace was 1 yard long. Approximately how fast, in miles per hour, was the jogger moving?

TEACHING RESOURCES

©2004 Key College Publishing, *Instructor Resources: Crossing the River with Dogs*, Johnson/Herr/Kysh

5. THE BRIDGE OVER RAT-TONGUE RAVINE

Sam Action had a tough task. He was a scoutmaster and had to get a group of Boy Scouts back to their campsite after a long hike. A number of these boys would whine and complain if they were left with some of the other boys unless Sam himself was present. The whining and complaining had something to do with which soccer team the boys played on and can be summarized as follows: Any of the six Rattlers would complain if left with any of the four Thunders. Either of the two Dragons would complain if left with the Eagle. Any of the Thunders would complain if left with the Eagle or one or both Dragons. Sam had to be present among any of the complainers to prevent complaining. All went well until they reached Rat-Tongue Ravine with the rickety bridge over it. Sam estimated that the bridge could safely handle only five people at a time, though he figured that it should be able to handle six people for three trips. He had been on the bridge before and knew it well. Sam needed to guide each group across to avoid capsizing the bridge, so he planned to cross the bridge with each group. He had to devise a way to get everyone across without any complaining. How did Sam arrange crossing the ravine in the least number of tries?

©2004 Key College Publishing, *Instructor Resources: Crossing the River with Dogs*, Johnson/Herr/Kysh

Problem Set B, Version 3

1. HOT DOG

If 30 hot dogs can feed a family of five for three meals, how many hot dogs would you need to feed just the three kids for eight meals? (Assume that adults and kids eat the same amount.)

2. FRATERNITY OUTING

The fraternity brothers at Phi Phi Pho Phum had arranged an outing with some other fraternities. They planned to attend a pie-eating contest together. There were 5 members of Phi Phi Pho Phum, 4 members of Eta Pi, 3 members of Nu Kappa, 2 members of Beta Zeta Theta, and 1 member from Tau Rho. They all met in front of the library to drive over to the contest. Unfortunately, they had only one vehicle: an old, beat-up Volkswagen "Bug." They figured it would hold 6 people plus the driver, who was John Vernon, dean of the fraternity council. The giant Phi Phi Pho Phum members were exceptions to this formula. Each of them took as much room in the car as 2 members of any other fraternity. To further complicate matters, many of the fraternities were fiercely competitive and would play practical jokes on some of the others, even if outnumbered. The competitors are as follows: The Tau Rho member would be quite eager to play a trick on anyone else from any of the other fraternities. Any of the Nu Kappa members would gladly play tricks on any of the Eta Pi members. Any of the Phi Phi Pho Phum members would gladly make fun of any of the Beta Zeta Theta members. No tricks would be played in the car while Vernon was present, and no tricks would be played when all 15 members were together at the original gathering or at the pie-eating contest. How did Vernon arrange the transportation in the least number of trips, so that no member played a trick on anyone else?

3. WHICH CAR SHOULD THEY TAKE?

The Jones family has two cars, a compact car and a large van. They can never decide which car to take when they visit the kids' grandmother. The kids like to take the van because it is much bigger and they have room to move around inside the van. The father likes to take the compact because he hates driving the van and the compact gets much better gas mileage. One day, they decided to find out how much they would save if they took the compact and whether it was worth the extra hassle. The compact gets 30 miles per gallon, and the van gets 20 miles per gallon. The kids' grandmother lives 2½ hours away when they drive 60 miles per hour. Gas costs $1.98 per gallon. How much more does it cost them to take the van?

©2004 Key College Publishing, *Instructor Resources: Crossing the River with Dogs*, Johnson/Herr/Kysh

Problem Set B, Version 3 (continued)

4. HOT DOGS AGAIN

My mother told us kids that if we wanted to go on this camping trip, we had to plan the meals. No problem. We planned to have hot dogs. That's not all, of course. We also took along mustard, relish, catsup, and buns. We figured we had it made. However, after setting up the tent, we didn't take care to make sure that the food was safe. Raccoons got into it and ate ⅓ of the hot dogs. After discovering the loss, our family had 6 hot dogs with our eggs for breakfast. When we went out sightseeing, we just left the hot dogs on the table and returned to find that jays had eaten another ¼ of our remaining supply. Despite our bad luck with the food, we continued to enjoy the camping trip. We went to the store and bought 12 hot dogs and added them to our supply. Then some of our campsite neighbors had a hankering for normal food and "borrowed" ⅖ of the remaining hot dogs. At our dinner that evening, we ate 10 hot dogs. That night, bears got ¾ of the last part of our supply, and in the morning we all shared the remaining 2 hot dogs. How many hot dogs did we start with?

5. LOST IN GRIDLOCK

Jack Holentyre was fed up with gridlock. He was stuck in some heavy traffic in the San Fernando Valley and abandoned his car (it was a piece of junk anyway). The unfortunate thing is that he left most of his navigating brains in his car. (The car knew how to get home; Jack didn't.) So Jack decided to "hoof it." He started out walking north. He walked 1 block north, then turned right and walked another block. Then he turned right again and walked another 2 blocks. He turned right again and walked 2 more blocks. By now he had decided on a pattern: Jack would keep increasing his distance after every two directions. (He had already walked 1 block north and then 1 block east. Then he increased his distance and walked 2 blocks south and 2 blocks west.) Next he would walk "3 and 3," as in 3 blocks north and 3 blocks east. Then "4 and 4," and so on. He figured that if he kept turning and kept changing his distance, eventually he'd get close to his house and recognize something. After he had walked 75 blocks in this fashion, he recognized his friend's house. He knew how to get home from there. He then walked 6 blocks north and 2 blocks east to his own house. Unfortunately, Jack left the envelope he needed to mail (with the car payment) back in his car. Help Jack out by giving him directions back to his car by a direct route.

©2004 Key College Publishing, *Instructor Resources: Crossing the River with Dogs*, Johnson/Herr/Kysh

Problem Set B, Version 4

1. LOST CARDS

My sister Allyndreth has really bad luck. She started a baseball card collection last year, but unfortunately, she can't seem to keep track of all of her cards. Recently she had a particularly bad string of luck. She took all of her cards with her on the bus to school. On the way there, some bullying sixth graders stole ¼ of her cards. Then she dropped six cards in the gutter when she got off the bus. She went to class and lost ⅔ of her remaining cards during show-and-tell. Then she gave three cards to her best friend. The teacher, Mr. Devlin, was really interested in baseball, so Allyndreth gave ⅗ of her remaining cards to the teacher. When she went to recess, she found eight cards on the playground. Then at lunch, she accidentally threw half of her remaining cards away in her lunch bag. When she came home, she only had ten cards left. How many cards did Allyndreth have when she left for school?

2. FENCE PAINTING

Three painters can paint six fences (all the same size) in three hours. How many fences can seven painters paint in five hours?

3. MARK AND DIANE

In this city, all the streets that run north and south have lettered names (A, B, C, etc.) and all the streets that run east-west have numbered names (1st, 2nd, 3rd, etc.). As you drive east, the letters get later in the alphabet. As you drive south, the numbers get bigger. Mark lives at the corner of A and 1st. His girlfriend, Diane, lives at the corner of E and 8th. Mark enjoys riding his bike to Diane's house. However, he likes to go a different way every day. If he only rides east and south, how many different routes can he take from his house to Diane's house?

4. WALKING

I was walking down the street following someone. He was at a pole as I passed a parking meter. It took me 14 seconds to get to the pole. When I reached the pole, he was at a fire hydrant. It took me 11 seconds to get to the hydrant. If I walk 3 miles per hour, how far will I have to walk from the parking meter to the point where I pass him?

Problem Set B, Version 4 (continued)

5. WEDDING RECEPTION

The long-awaited marriage of Amanda Smith and Kevin Jones had finally taken place. But trouble was brewing at the reception. A horse and buggy had been hired to take the most-honored guests and some members of the wedding party from the church to the reception. The following people needed to be transported in this buggy, which held at most four people: the bride and groom, Amanda Smith and Kevin Jones; Phil and Jerri Smith, parents of the bride; Len Smith, brother of Phil; Walter Lewis, brother of Jerri; Lincoln and Jill Jones, parents of the groom; J. R. Jones, brother of Lincoln; J. R.'s wife, Mina; and Brian Jones, son of Mina and J. R. Unfortunately, the transportation had the following restrictions: Jerri must go on every trip, as she is the mother of the bride and hostess of the wedding. Married couples must always remain together (this restriction does not include Jerri and Phil, as they are divorced). Len must always be with Phil, as Phil is the only one who will laugh at Len's jokes. Brian cannot be left at the same location as or ride with Amanda, because sparks tend to fly when they get together. Brian can't be left with or ride with Walter either, because Brian tends to talk about the Miami Dolphins and Walter hates the Dolphins. Jill cannot be left with or ride with J. R., because they used to be married but Jill divorced J. R. and married his brother, Lincoln. How can the transportation to the reception be arranged in the least number of trips? (*Note:* Restrictions do not apply at the church *before* the first trip leaves, and they don't apply once everyone has arrived at the reception.)

©2004 Key College Publishing, *Instructor Resources: Crossing the River with Dogs*, Johnson/Herr/Kysh

12 Draw Venn Diagrams

V enn diagrams allow you to organize information spatially. In particular, they make classification schemes easier, especially complex classifications that cannot be well represented linearly, such as in a list.

Make sure students understand the three basic relationships of Venn diagrams. The first is the disjoint relationship, depicted with two non-overlapping loops. The objects represented by one loop have nothing to do with the objects represented by the other. The word *no* often appears in descriptions of disjoint relationships, as in the statement "No lawn mowers are roses."

 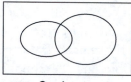 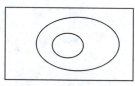

Disjoint   *Overlapping*   *Subset (enclosed)*

In overlapping loops, the overlap represents objects with characteristics common to two sets of objects. The word *some* often describes overlapping relationships like "Some math students are chemistry students." Finally, when one loop encloses another entirely, the inner loop represents a subset of the objects represented by the outer loop. Inner loops possess all the characteristics represented by any subsequent outer loop. For example, "All roses are flowers" states that roses are a subset of flowers and possess all the characteristics of flowers.

The universal set, usually drawn as a rectangle, describes the most inclusive set among those under consideration. Animals would be a universal set encompassing household pets, cats, and dogs.

It is important to lay some ground rules for students. You can say a number of things about categories of animals, such as "No bears are lawn mowers." Some students may propose a pet bear whose name is "Lawn Mower." Emphasize that we are using the principal, common understanding of this statement.

Negations, that is, attributes that objects don't possess, are key and should be made clear. Many times we label the different zones in Venn diagrams and then describe the things represented by those zones.

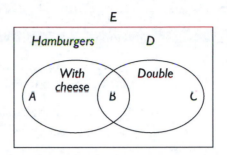

Students should give the most specific characterization possible for each zone. The list should include attributes the objects in that zone do and don't have.

A: Hamburger, with cheese, not double

B: Hamburger, with cheese, double (a double cheeseburger)

C: Hamburger, double, no cheese

D: Hamburger, no cheese, not double

E: Not a hamburger

Prepare students for Venn diagrams by having them classify themselves in a Venn diagram on the board. Draw a Venn diagram on the board like the following diagram. As students enter, they write their initials in the appropriate region. Make it clear to students that they can initial in only one place. Venn diagrams are often extremely intuitive, so students can understand the concepts by relying on their previous experience.

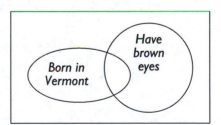

©2004 Key College Publishing, *Instructor Resources: Crossing the River with Dogs*, Johnson/Herr/Kysh

Here are other examples of good Venn diagrams to use on the board. We like to put these and many others on the board and have students walk around the room signing their name into each diagram.

1. Four concentric circles labeled with the country, state or province, county, and city your school is in. Ask students to initial the region that describes where they were born.

2. Three intersecting circles labeled with three sports. Ask which sport or sports students like to play or watch.

3. Three intersecting circles labeled with popular TV shows. Ask which show or shows students like to watch.

Students may have trouble counting correctly. The diagram below represents ten flowers. Six of the ten are roses. Four of the ten flowers are not roses. Students need to be careful that when they write a number in a region, the number represents only objects in that region, not objects in subregions. (In this example, the four flowers obviously do not include the six roses.)

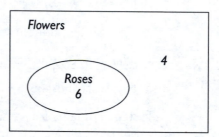

You can go outside for a fun activity that teaches Venn diagrams. Draw any sort of looping relationships with chalk (or use rope or hoses). You can tell students what the loops represent and let them find the appropriate region to stand in. Better yet, keep the loop relationships to yourself, then direct students to the appropriate regions and leave it to them to figure out what the regions represent.

Are Venn diagrams used in real life? Yes! They are frequently used in scientific classifications, sociology, political science, and information science. Computerized searches can be modeled with Venn diagrams. If there are listings for 1200 articles on the depletion of the ozone layer and 800 articles on genetic mutation, by asking for the articles that have both characteristics you should get a smaller listing that is a subset of both groups. By including more categories or by including more-specific categories, the numbers should get even smaller. This is an example from political science:

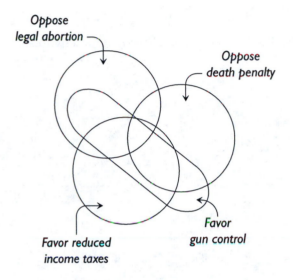

A person who fits into any one of these categories might be described as conservative or liberal. But there are 15 different regions in the diagram (16 if you count the region outside of any loop). Making the Venn diagram helps reveal the complexity of the relationships.

An alternative way to make a diagram that includes four categories is to use a grid:

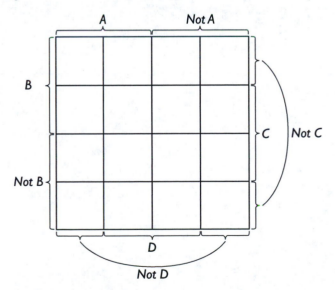

Good Quiz Problems

For a one-problem quiz on this strategy, consider one of these problems from Problem Set A, Version 2:

2. Soda Survey

3. Company Barbecue

5. Country Club

10. Marketing Company Does Soda Survey

11. Health Club Activities

TEACHING RESOURCES

©2004 Key College Publishing, *Instructor Resources: Crossing the River with Dogs*, Johnson/Herr/Kysh

Text Problems

BASIC RELATIONSHIPS

Your understanding of creating accurate Venn diagrams is critical to understanding the problems in the rest of this chapter. Draw Venn diagrams that represent each of the following statements:

1. Some birds are pets.

2. No dogs are sheep.

3. All poodles are dogs.

4. Some dogs are poodles.

Draw these diagrams before continuing.

©2004 Key College Publishing, *Instructor Resources: Crossing the River with Dogs,* Johnson/Herr/Kysh

CATEGORIES

For each list of categories, draw a Venn diagram that shows the relationships among the categories. One of the categories describes the universal set, and the others describe the various loops inside the universal set. The universal set is not necessarily listed first.

1. Household pets, dogs, animals, cats

2. Living things, lizards, apes, chimpanzees, reptiles, dogs, mammals, terriers, dachshunds

3. Place of birth, USA, Canada, Miami, Florida, Orlando, Montreal, Missouri

4. Universities, private universities, public universities, Yale, University of Texas at El Paso (UTEP), Notre Dame

5. Hamburgers, with cheese, double, homemade

Work this problem before continuing.

©2004 Key College Publishing, *Instructor Resources: Crossing the River with Dogs,* Johnson/Herr/Kysh

SCIENCE COURSES

In a group of students, 12 are taking chemistry, 10 are taking physics, 3 are taking both chemistry and physics, and 5 are taking neither chemistry nor physics. How many students are in the group? Work this problem before continuing.

©2004 Key College Publishing, *Instructor Resources: Crossing the River with Dogs,* Johnson/Herr/Kysh

©2004 Key College Publishing, *Instructor Resources: Crossing the River with Dogs,* Johnson/Herr/Kysh

Text Problems *(continued)*

MUSIC SURVEY

In a poll of 46 students, 23 liked rap music, 24 liked rock music, and 19 liked country music. Of all the students, 12 liked rap and country, 13 liked rap and rock, and 14 liked country and rock. Of those students, 9 liked all three types of music. How many students did not like any of these types? Work this problem before continuing.

SPRING ROSTER

A manager of a baseball team looked over his roster at the beginning of spring training. He noticed the following facts: Every outfielder is a switch hitter. Half of all infielders are switch hitters. Half of all switch hitters are outfielders. There are 14 infielders and 8 outfielders. No infielder is an outfielder.

How many switch hitters are neither outfielders nor infielders? Work this problem before continuing.

Problem Set A, Version 2

1. FORMAL DANCE

There are 120 students in the dorm. Of those, 66 are going to the formal dance. The same number (66) of students are going out to dinner that evening, including 48 of those going to the dance. How many students are neither going to the dance nor going out to dinner?

2. SODA SURVEY

In a recent consumer interview, 800 people took taste tests. Of those, 723 said they like soda. Of the people who like soda, 316 like both root beer and cola, while another 87 said they like cola but do not like root beer. Furthermore, a total of 427 people in the survey like root beer. How many people in this survey like soda but do not like either root beer or cola?

3. COMPANY BARBECUE

Before doing the ordering for a company barbecue, the president took a poll to find out how many people would eat ribs, hamburgers, and/or corn at the barbecue. She polled everyone in the company who said they would probably attend. This poll revealed that 57 people would eat ribs, 38 would eat corn, 65 would eat hamburgers, 30 would eat ribs and corn, 24 would eat corn and hamburgers, 40 would eat ribs and hamburgers, 19 would eat all three, and 7 wouldn't eat any of these three foods.

Draw the Venn diagram. Then answer these questions:

a. How many people were surveyed?

b. How many people would not eat ribs?

c. How many people would eat ribs but not corn?

d. How many people would eat corn and hamburgers but not ribs?

e. How many people would eat only hamburgers?

4. SPIRIT OF WOODSTOCK

Of the six members in the Spirit of Woodstock rock band, four can play guitar. There are three who can play keyboards. All of the singers play guitar, and two of the guitarists also play keyboards. Two members do all three. One guitarist doesn't sing. How many members sing but do not play keyboards?

©2004 Key College Publishing, *Instructor Resources: Crossing the River with Dogs*, Johnson/Herr/Kysh

Problem Set A, Version 2 *(continued)*

5. COUNTRY CLUB

A survey was given among the members of the Silver Springs Country Club, asking which activities each club member participated in. The survey revealed that 67 played tennis, 64 played golf, and 52 liked to swim. Of those, 29 liked to play golf and tennis, 27 liked tennis and swimming, 31 liked to play golf and swim, and 14 participated in all three. There were 5 members who did not do any of the activities. How many members were surveyed?

6. EAST PARKING LOT

On the night our play-off was being held, a number of vehicles were in the east parking lot of the gym. Half of the white vehicles were neither cars nor buses. There were 8 buses in all, and only 1 of those was yellow. There were 16 other yellow vehicles, though, and 6 of those were cars. Of a total of 20 cars, 9 were not yellow or white. There were as many white buses as there were buses that were not white. Besides cars and buses, of course, there were vans and trucks. How many white vehicles were there?

7. SPLIT TICKET

In the fourth Dodgeville precinct, the Democrats and Republicans both had one candidate each for the House of Representatives, for the Senate, and for president. There was, however, one additional candidate for Congress, who was registered as a Democrat and running a write-in campaign. In the voting, one-seventh of registered voters did not vote. Seven wrote in the congressional candidate and voted the rest straight Democratic (voted for the one write-in candidate for Congress, the Democratic candidate for senator, and the Democratic candidate for president). Forty-one people voted for at least one Republican candidate, though none of those were the ones who voted for the write-in candidate. Thirty-seven voted for at least one Democrat, including the 18 who split their ticket between Democrats and Republicans. How many people were registered in the fourth precinct?

8. BRAVE ALL-STARS

A number of people in the Atlanta Braves organization got together recently for a party. Of those, $1/12$ were all-stars; $2/3$ of the all-stars were hitters. Of the people at the party, $1/2$ were players, and $1/2$ of the players were neither hitters nor all-stars. Of the hitters, $1/4$ were all-stars. There were 6 hitters who were not all-stars. How many people were at the party?

©2004 Key College Publishing, *Instructor Resources: Crossing the River with Dogs,* Johnson/Herr/Kysh

Problem Set A, Version 2 *(continued)*

9. EARTHQUAKE DAMAGES

After an earthquake, 32 homes in a remote area were checked by public safety officers before nightfall. They found that 22 of the houses had no gas, including 10 houses whose own gas lines were damaged. (A house loses its gas service if it has damage to its own line or if there is damage to the feeder line.) Of those houses that lost their gas, only 4 did not lose their electricity. Twenty-four houses in all had no electricity; the same number of them lost only their electricity as those that both lost electricity and had damage to their gas lines. How many of the houses checked did not lose either gas or electricity, and how many lost only their gas service?

10. MARKETING COMPANY DOES SODA SURVEY

A marketing company did a survey to find out what kinds of soda people drink. The survey found that 59 drink cola, 43 drink root beer, and 51 drink lemon-lime. Of those, 23 drink cola and root beer, 35 drink cola and lemon-lime, 22 drink root beer and lemon-lime, 15 drink all three kinds of soda, and 2 drink none of these three kinds of soda.

  a. How many people were surveyed?

  b. How many people do not drink cola?

  c. How many people drink root beer but not lemon-lime?

  d. How many people drink cola and lemon-lime but not root beer?

  e. How many people drink only root beer?

11. HEALTH CLUB ACTIVITIES

A survey was given to the members of the Muscle Bound Health Club, asking which activities each club member participates in. The survey revealed that 41 like aerobics classes, 31 like the treadmill, and 38 like lifting weights. Of those, 24 like aerobics and the treadmill, 22 like aerobics and weights, 19 like the treadmill and weights, and 14 like all three activities. There were 4 members who do not like any of these three activities.

  a. How many members were surveyed?

  b. How many people like weights but not aerobics?

  c. How many people like aerobics and the treadmill, but don't like weights?

  d. How many people don't like the treadmill?

  e. How many people like only aerobics and nothing else?

©2004 Key College Publishing, *Instructor Resources: Crossing the River with Dogs,* Johnson/Herr/Kysh

Problem Set B, Version 2

1. THE TREASURE OF SIERRA MARBLES

Tom Sawyer, Huck Finn, and Becky Thatcher were feeling pleased with themselves. They had just found a stash of great marbles buried in Tom's backyard in the Sierra Nevada mountains (they had moved from Missouri)—left, no doubt, by some previous tenant. They decided to sleep in the backyard that night to guard the marbles from theft. Unfortunately, they should have been guarding the marbles from each other. Soon after everyone else had gone to sleep, Tom got up and decided to take his fair share of marbles. He separated the marbles into three piles. There was 1 marble left over, which he buried, and then he took one of the piles of marbles, put the remaining marbles back into one big pile, and went back to bed. A little while later, Huck got up and did the same thing. He separated the marbles into three piles. There was 1 left over, which he buried. He took one of the three piles, combined the other two, and went back to bed. Becky got up last and found 2 marbles half buried in the ground. She put them with the remaining pile of marbles. Then she took one-third of the marbles that remained (the remainder was zero when she divvied them up) and went back to bed. In the morning the three friends got up and looked at the small pile of marbles that remained. An argument immediately broke out, as each friend accused the others of taking some of the marbles. No one would admit guilt, so they ended up splitting the remaining marbles evenly, which came to 16 more marbles per person. How many marbles did each person end up with?

2. TOOTHPICK SQUARES

What is the smallest number of toothpicks necessary to make 100 little squares? All squares are to be the same size. Toothpicks do not overlap. Discuss how you can know that your arrangement uses the fewest toothpicks. For example, to make 5 squares you need either 15 or 16 toothpicks as shown.

©2004 Key College Publishing, *Instructor Resources: Crossing the River with Dogs*, Johnson/Herr/Kysh

Problem Set B, Version 2 (continued)

3. THE FRESHMAN CLASS

Ten students are taking neither general psychology nor elementary Spanish. Five students are taking neither linear algebra nor Spanish. Seven students are taking neither algebra nor psychology. All of the students are taking at least one of the three subjects. No student is taking all three subjects. Twenty-one students are not taking psychology. Seventeen students are taking algebra. Seventeen students are taking Spanish. How many students are taking psychology?

4. THE SANDWICH SHOP

Several teachers invested together in a sandwich and soda shop. They call it the Teacher's Sub Shop. They offer custom-made sandwiches. You can buy a sandwich with the following bases: pastrami, turkey, roast beef, or vegetarian. Along with that, they offer four kinds of cheese: monterey jack, cheddar, swiss, and american. You can also have mustard, mayonnaise, both, or neither. The patrons also get to choose white, wheat, or rye bread, but the rye is not offered with either pastrami or cheddar, as one of the owners objects under the rationale that the combinations of cheddar and rye or pastrami and rye are "gross." You also can't get mustard on turkey or mayonnaise on roast beef. A sandwich consists of one base, one cheese if desired, condiments if desired, and one bread. How many different types of sandwiches does this shop offer?

©2004 Key College Publishing, *Instructor Resources: Crossing the River with Dogs*, Johnson/Herr/Kysh

Problem Set B, Version 2 *(continued)*

5. FINE ARTS

Five people from Nashville recently ordered tickets to their favorite fine arts events: jazz concert, art exhibit, symphony concert, ballet, and musical comedy. None of the five people like to go to the same event. Their first names are Alan, Bev, Chris, Doreen, and Ernie, and their last names (in no special order) are Fillmore, Gunderson, Hatfield, Innis, and Jackson. The occupations of the five are publisher, engineer, reporter, doctor, and lawyer. Use this information and the clues below to determine each person's full name, occupation, and favorite fine arts event.

1. The kids of Jackson, Innis, and the person who is going to the jazz concert all play on the same soccer team.

2. Among the lawyer's clients are Alan and the art enthusiast.

3. Doreen and the reporter and the person who likes the symphony all have their tickets already.

4. Alan, Hatfield, and the jazz fan all attended the first Tennessee Titans football game together.

5. Neither Ernie nor the reporter has kids.

6. Of the five, only the doctor has never attended the symphony. The engineer, who is not Chris, has attended ballets and art exhibits, but didn't like either.

7. The person who enjoys musical comedy (especially musicals written by Stephen Sondheim) and the doctor both have their tickets and are going to attend their events in the next few days.

8. The doctor (who is not Innis) and Ernie (who has never heard of Stephen Sondheim) went fishing the day of the first Tennessee Titans football game because both of them hate football.

9. Fillmore and the engineer just recently ordered their tickets and have not received them yet.

TEACHING RESOURCES

©2004 Key College Publishing, *Instructor Resources: Crossing the River with Dogs*, Johnson/Herr/Kysh

Problem Set B, Version 3

POLITICAL PARTIES

Five friends who belong to the same political party have known each other for several years. Like most friends, they have a lot in common and a lot of differences. As they were talking at a rally, they noticed some things that they all did differently. Each goes to work in a different way: There is a car driver, a biker, a walker, a subway rider, and a bus rider. Their first names, not in any special order, are Walter, Teresa, Valerie, Jasper, and Maria. Their last names are Doyle, Peterson, Sinderson, Archer, and Lamson. And although they all are political activists, each one votes in a different area: Queens, the Bronx, Manhattan, Long Island, and Seaside. From the clues below, determine each person's full name, the way they get to work, and where they vote.

1. Teresa, the biker, and the person who votes in Manhattan have all sent in their absentee votes. Archer and the person who takes the bus haven't voted yet.

2. The person who votes in the Bronx, Doyle, and Sinderson have all held office before.

3. The subway rider and the car driver meet Doyle and the person who votes in the Bronx for lunch every Tuesday. Walter cannot attend because he works on the other side of the city.

4. The person who votes in Queens and the subway rider voted absentee. They discussed their votes with the woman from Seaside (who hasn't voted yet).

5. Neither Valerie nor the person who bikes to work has ever held office.

6. Jasper (whose last name is not Sinderson) and Mr. Peterson and the person who votes in the Bronx first met at a victory party ten years ago.

7. Maria could take the subway but prefers her normal mode of transportation.

8. The person who rides a bus to work does not live in Long Island or Seaside.

2. **COUNTRY MUSIC**

Maile was driving through the Midwest recently and happened to tune in to a country music station. She likes country music, so she really enjoyed getting a chance to listen to some. After having listened to the station for an hour, Maile had heard 25 songs. All but 1 song were about either truck drivers, being in love, hound-dog owners, or some combination of the three. She noted the following information:

All of the truck drivers were in love. Three-fourths of the truck drivers were not hound-dog owners. There were six hound-dog owners who were in love. Of the hound-dog owners who were not truck drivers, half of them were in love. Eighty percent of all the songs concerned people in love. How many songs concerned people who were in love but were not truck drivers or not hound-dog owners?

3. **HUNGRY BROTHERS**

Four hungry brothers went to bed, thinking about the batch of cookies that their mother had just baked. In the middle of the night, each of them got up and ate some of the cookies. Allen got up first. He separated the cookies into three piles and found there was 1 left over. He gave the leftover cookie to the dog, then ate 1/3 of the remaining cookies and went back to bed. Brian was the next to get up, and he ate 1/4 of the cookies. (The dog didn't get any from him because he thought the dog looked full.) Colin then got up and gave 2 cookies to the dog. Then he ate 1/5 of the remaining cookies. Devon was the last to get up, and he ate just 4 cookies and then gave 1 to the dog, who was beginning to look sick. When their mother, Vickie, went into the kitchen the next morning, she found a lot of crumbs, ants, a pale-looking dog, and only 15 cookies. How many cookies had she baked?

4. **LET GO MY LEGO**

Cody built a pyramid out of Legos. He used rectangular Legos that were four bumps by two bumps. He started with one Lego in the top level. Underneath that he put in as many Legos as needed to leave one row of bumps all the way around the outside. He continued to build the pyramid in such a way that there was always one row of bumps all the way around the outside on each level. He was able to make ten layers on his pyramid before he ran out of Legos. The inside of the pyramid is completely filled in with Legos, many of which are not visible to someone looking at the top of the pyramid.

How many Legos are in the pyramid?

How many bumps total on all levels are visible on the outside of the pyramid?

©2004 Key College Publishing, *Instructor Resources: Crossing the River with Dogs*, Johnson/Herr/Kysh

Problem Set B, Version 3 *(continued)*

5. **HOW MUCH STEREO CAN YOU AFFORD?**

You are interested in buying a stereo system for your college dorm room. You want to buy a receiver, a compact disc player, and a pair of small bookshelf speakers. You consult a consumer magazine and determine that you are interested in several kinds of each item. The list is shown below with prices.

CD PLAYERS	RECEIVERS	SPEAKERS (PAIR)
$400 Megavox	$300 Yohaha	$450 Allinfun
$330 BLT	$400 Technical	$350 Infinitesimal
$500 Bashiba	$250 Shark	$690 Mandarin
$550 Denmom	$490 Pilgrim	$200 ERP
	$350 Oinko	$300 Scoot
		$500 Mansion

After visiting your friendly neighborhood stereo store, you find it is out of the BLT CD player, and you don't like the Yohaha receiver or the Mansion speakers. You have $1,100 to spend on your stereo system. In how many different ways can you buy a system that you will be happy with? (You do not have to spend all of the money.)

Problem Set B, Version 4

1. FOUR FRIENDS

Four friends (one is named Sara) drove to the grocery store to buy their favorite food and soda. Two sat in the front seat, and two sat in the back seat. Each bought a different food (one bought chili), and each bought a different soda (one bought root beer). Use the clues below to determine what food each person bought, each person's soda flavor, and where each person was sitting.

1. The person who bought soup sat next to Marco.

2. Roberto sat diagonally from the person who had the lemon-lime soda.

3. The person who bought cola sat on the same side of the car as the person who preferred macaroni and cheese.

4. The person who had the orange soda sat on the same side of the car as the person who bought soup.

5. Roberto sat on the same side of the car as the person who bought spaghetti.

6. Areatha sat next to the person with the cola.

7. The driver did not buy a kind of pasta.

8. The person with the root beer sat in the back seat.

2. SOCCER TEAM

On a soccer team there are four positions: goalie, back, midfielder, and forward. Each player on the team often plays more than one position during the course of the season. On one soccer team there are 16 members. All of the goalies also play forward, and one goalie plays back as well. No goalie plays midfielder. There are as many backs as midfielders. The total number of midfielders is two-thirds the total number of forwards. Half of the team plays midfielder. No one plays only back. Three people play three positions. Three people play only one position. The number of midfielders who play forward but not back is equal to the number of midfielders who play back but not forward and is equal to the number of midfielders who don't play anything else.

How many people play goalie? How many people play back and forward but nothing else?

Problem Set B, Version 4 *(continued)*

3. ORANGE TREASURE

Three pirates were stranded on a desert island just outside of Pittsburgh. They came upon a treasure chest, which they opened and found to be full of oranges. Imagine their surprise. These pirates were overjoyed, however, because they loved oranges. Unfortunately, they couldn't decide how to divide them up, so they left them in the chest. On the first afternoon, one pirate, Ahab, snuck away from the others and went to the treasure chest. He separated the oranges into three equal piles. There was 1 extra orange, which he gave to a passing monkey. Then he ate one of the three equal piles of oranges, put the rest back into the chest, and went back to camp. A short time later a second pirate, Bluebeard, snuck away from camp and went to the chest. He separated the oranges into four equal piles. (He wasn't quite as greedy as Ahab.) There were 2 extra oranges, which he gave to a passing monkey. He then ate one of the four piles and put the rest back into the chest. Still later the third pirate, Hook, went to the chest and separated the oranges into five equal piles. (He was even less greedy.) There were 3 left over, which he gave to a passing monkey. Then he ate one of the piles and put the rest back into the chest. The next day the three pirates split the remaining oranges equally. Each pirate got 20 oranges. How many oranges were in the chest originally, and how many did each pirate receive?

4. PENTAGON

If all the diagonals of a regular pentagon are drawn, how many triangles are formed? (*Regular* means all sides and all angles are the same.)

5. DIGIT 8

If all of the numbers from 1 to 999,999 were written down, how many times would the digit 8 appear in the list? (Examples: 38 has one 8; 35,828 has two 8's; 88,348 has three 8's; 345,921 has zero 8's.)

©2004 Key College Publishing, *Instructor Resources: Crossing the River with Dogs,* Johnson/Herr/Kysh

13 Convert to Algebra

lgebra is an outrageously powerful problem-solving strategy—so powerful that it has monopolized the field for years, to the detriment of other problem-solving strategies and to algebra itself. Our intent with this chapter is not to teach algebra; we know a few books that already do that, and we recommend algebra as a prerequisite for students taking this course. Instead, we use the chapter to demonstrate how other problem-solving strategies, especially guess-and-check, can be used in conjunction with algebra.

Many current algebra textbooks include a page on guess-and-check. The page may include "coin" problems, "age" problems, "work" problems, and a few other types. Entire lessons in the book might be devoted to these problem types, but guess-and-check is treated as a sidelight, without much explanation of how it can be applied to these problems.

In *Crossing the River with Dogs*, when algebra is used with guess-and-check, algebra becomes more powerful and accessible. Students understand guess-and-check better than they do algebra, so using guess-and-check to develop algebraic equations helps transfer this understanding to the algebra.

The biggest weakness of guess-and-check is in the handling of fractional answers. It is difficult to guess and check when the answer you're looking for is $211/315$ and the closest you can come is $322/480$. One way around this is to determine to what decimal place you want the answer. Answers to the nearest tenth are often reasonable. For better precision at that point, you should probably convert the guess-and-check procedure to an algebraic equation.

Virtually any problem in an algebra text can also be solved with guess-and-check. Students who successfully solve a problem with guess-and-check should also be able to set up an equation. They use a variable for the guess. For each column in the chart, they apply to the variable the operations they applied to their previous guesses. They should think, "What did we do to get this column?" to come up with an algebraic expression and write the result in that column. The equation is usually right there in one or two of the columns. The only thing left to do is to figure out what column they've been doing the check on. That column often contains both sides of the equation. One side is the result of the algebraic guess. The other side is what the result of the guess is supposed to equal. If they've been trying to make

TEACHING RESOURCES

©2004 Key College Publishing, *Instructor Resources: Crossing the River with Dogs*, Johnson/Herr/Kysh

two columns equal each other, those columns most likely contain the two sides of the equation.

Guess-and-check takes practice, just as writing and solving equations do. Students might solve some problems in this chapter by guess-and-check alone, but require them to write algebraic guesses anyway so that they get enough experience with this powerful strategy.

On a practical note, point out to students that someday in the future they may be called upon to do some algebra on the job. Even if they've forgotten their algebra, guess-and-check can be used in many instances and can help them remember how to set up equations for problems with non-integer answers.

Note on a Text Problem

 SATURDAY AT THE FIVE-AND-DIME GARAGE SALE
This problem, like many others, can be set up algebraically in either one or two variables.

Good Quiz Problems

Many of the problems in Problem Set A, Version 2, would make good quiz problems. Depending on how much you want to emphasize algebra, you may wish to give a quiz of from one to four problems. Make every problem a different type. These problems do take students a long time to do, so bear that in mind. Good quiz problems are

1. More and More Coins
3. MaryLou's Investments
9. Terry's Boo-Boo
14. Sugary Punch
17. Another Coin Problem
18. Still More Coins
19. Decent Investments
20. Investments
21. Visiting a Friend
22. Road Trip
23. Rachael's Juice
24. Talia's Truly Terrific Fruit Punch

Text Problems

SATURDAY AT THE FIVE-AND-DIME GARAGE SALE

Cinci held a garage sale, during which she charged a dime for everything but accepted a nickel if the buyer bargained well. At the end of the day she realized she had sold all 12 items and had a total of 12 nickels and dimes. She had raked in a grand total of 95 cents. How many of each coin did she have?

See if you can re-create the guess-and-check chart for this problem without looking back at Chapter 6.

FARMER JONES

Farmer Jones raises ducks and cows. She tries not to clutter her mind with too many details, but she does think it's important to remember how many animals she has and how many feet those animals have. She thinks she remembers having 54 animals with 122 feet. How many of each type of animal does Farmer Jones have? Again, re-create the guess-and-check chart for this problem. Then set up the equation or equations to solve the problem algebraically.

ALL AROUND THE PLAYING FIELD

The perimeter of a rectangular playing field measures 504 yards. Its length is 6 yards shorter than twice its width. What is its area? Set up the guess-and-check chart for this problem, and then write the equation.

ORIGAMI

A group of exchange students from Japan went to a convalescent home to sing songs for the seniors and to demonstrate origami, the art of Japanese paper folding. Groups of students and seniors sat together at tables so the students could teach the seniors to fold origami models. As it turned out, at each table there was either 1 student at a table with 3 seniors, or 2 students at a table with 4 seniors. There were 23 students and 61 seniors in all. How many tables were being used? Work this problem before continuing.

Text Problems *(continued)*

CASCADES STATE PARK

Emi and Margit had stopped at the bottom of one of the highest waterfalls in Cascades State Park. As Emi looked up at the waterfall, she said, "Wow, I think the top of that fall is about 20 feet more than three times the height of that young redwood!" Margit, of course, had a different opinion. She said, "No, I think it's about 50 feet less than four times the height of the redwood." If both are approximately right, about how tall is the redwood and how high is the waterfall? Solve this problem by using a guess-and-check chart to set up the algebra before continuing.

CHOCOLATE MILK

Augustus is trying to make chocolate milk. He has made a 10% chocolate milk solution (this means that the solution is 10% chocolate and 90% milk). He has also made a 25% chocolate milk solution. Unfortunately, the 10% solution is too weak, and the 25% solution is way too chocolaty. He has a whole lot of the 10% solution, but he has only 30 gallons of the 25% solution. How many gallons of 10% solution should he add to the 25% solution to make a mixture that is 15% chocolate? (Augustus is sure the 15% solution will be absolutely perfect.) Work this problem before continuing.

SALT SOLUTION

A pet store sells salt water for fish tanks. Unfortunately, recently hired Flounder has mixed a salt solution that is too weak. He's made 150 pounds of 4% salt solution. The boss wants a 7% salt solution. Help Flounder out by giving him two options for reaching the 7% solution:

a. Add some salt. How much?

b. Evaporate some water. How much?

Work this problem before continuing.

THE SHADOW

A man 6 feet tall is walking away from a streetlight that is 15 feet tall. How long is the man's shadow when he is 10 feet away from the light?

Work this problem before continuing. Note that guess-and-check doesn't really help here, but a diagram helps a lot.

Problem Set A, Version 2

1. MORE AND MORE COINS

Rufus has $3.45 in quarters and dimes. He has four more quarters than dimes. How many of each coin does he have?

2. COMPLEMENTS

The larger of two complementary angles is 3 degrees more than twice the smaller of the two angles. What is the measure of each angle?

3. MARYLOU'S INVESTMENTS

MaryLou made some money on her investments. She made twice as much money on her investment at 8% than on her 5% investment. In all, she made $600 in one year. How much did she have invested at each percentage rate?

4. A SMALL WOODWORKING COMPANY

A Small Woodworking Company must ship two different-sized packages: The small ones cost them 45 cents each and weigh 6 ounces, and the large ones weigh 25 ounces and cost $1.20 each. The total shipment this morning weighed 20 pounds 7 ounces and cost $18.45. How many packages of each size were shipped?

5. SPEEDING TIX

The fines for speeding tickets inside the Kreith Township are computed as follows: The speed you were going is added to how much over the speed limit that speed is. The sum is then squared. The answer is the fine. Chester's fine for going 23 miles per hour over the posted speed limit was $5,041. How fast was he going and what was the speed limit?

6. DAD GETS LOTS OF HELP

Dad and Junior together can wash the car in 40 minutes. Junior can wash it by himself in 70 minutes. How long would it take Dad to do it by himself?

7. MOM GETS HELP?

Mom can rake all the leaves from the vast front yard in about 30 minutes. With help from her youngest child, Sonny, she can rake the yard in about 40 minutes. How much work does Sonny contribute to getting the job done? How long would it take Sonny working by himself?

Problem Set A, Version 2 *(continued)*

8. TAXING BERNICE

Bernice paid $2,970 in state income tax last year. The first part of her income was taxed at the rate of 7%, and the next (higher) part of her income was taxed at 9%. She had $42,000 in taxable income. How much was taxed at 7%, and how much was taxed at 9%?

9. TERRY'S BOO-BOO

Terry goofed up mixing 10 gallons of grape juice. He made a punch mix that was 20% juice concentrate instead of the usual 30% punch concentrate. How much juice concentrate does he need to add to the 10 gallons already mixed to bring it up to 30% juice concentrate?

10. ALGEBRA AREA

A rectangle with perimeter 104 centimeters has a width 18 centimeters less than its length. What is its area?

11. MICHAELA'S SEASON

Michaela had 24 hits last baseball season. She had no home runs but still managed to have 35 total bases out of those 24 hits. (A single counts for 1 base, a double counts for 2, and a triple counts for 3.) She had 7 doubles and triples total. How many each of singles, doubles, and triples did she have?

12. GOLD COUNTRY TIRE

Darcy runs a tire store. One week she counted 35 bills of sale. She knew that all of the customers had bought either 4 new tires or 2 new tires. She had 82 old tires to dispose of in the yard. How many customers bought 2 tires and how many bought 4 tires?

13. CLARENCE AND STEPHANIE ARE THINKING

Clarence and Stephanie are each thinking of a number. If you multiply Stephanie's number by 3 and subtract 2, you get Clarence's number. If you take half of Clarence's number and then subtract 2, you get Stephanie's number. What number is each thinking of?

14. SUGARY PUNCH

Mindy is mixing some sugary punch for tomorrow's fund-raiser. She took over for someone who goofed up. She knows that she already has 20 liters of an 18% punch mix and 50 liters of a 30% punch mix. How can she make both batches into a 25% mixture by mixing the existing punches and adding only water?

Problem Set A, Version 2 *(continued)*

15. SLED RUN

Vanessa and Kelsey set up a sled run for themselves and their friends. The first part dropped vertically about 16 feet over an 80-foot horizontal distance. The second part was flat for about 60 feet. The third part was about 150 feet horizontally with a 22-foot drop. About how far will the sled travel on this course?

16. FEDERAL STANDARDS

Federal standards require Gigantic Motors to sell cars that average 27.5 miles per gallon. If the fleet average was 25 miles per gallon for the first 150,000 cars they sold, how many cars averaging 35 miles per gallon must they sell to bring the fleet average up to 27.5 miles per gallon?

17. ANOTHER COIN PROBLEM

Satomi has $4.85 in dimes and nickels. She has a total of 60 coins. How many of each coin does Satomi have?

18. STILL MORE COINS

Joette has $16.95 in dimes and quarters. The number of quarters is three more than twice the number of dimes. How many of each coin does Joette have?

19. DECENT INVESTMENTS

Leonard invested money in two bank accounts. One account pays 2.9% annual interest, and the other account pays 6.15% annual interest. He invested four times as much money in the 6.15% account as he did in the 2.9% account. At the end of one year, Leonard made $345 in interest. How much money did he invest in the 6.15% account?

20. INVESTMENTS

Dara invested $7,000 in two bank accounts. One account pays 1.7% annual interest, and the other account pays 4.25% annual interest. At the end of one year, Dara made $189 in interest. How much money did she invest in the 1.7% account?

21. VISITING A FRIEND

Adri rode her bike to Alisha's house. On the way there, she rode 16 miles per hour. On the way back, she rode 12 miles per hour and it took 45 minutes longer. How far is it from Adri's house to Alisha's house? (*Hint:* How many hours is 45 minutes?)

Problem Set A, Version 2 *(continued)*

22. ROAD TRIP

Luz and Larry took a long road trip. They started their trip in Lodi. They drove at a speed of 58 miles per hour from Lodi to Merced. Then they drove at a speed of 62 miles per hour from Merced to Bakersfield. The trip from Merced to Bakersfield took two hours longer than the trip from Lodi to Merced. The total distance from Lodi to Bakersfield is 298 miles. How many hours was the trip from Lodi to Merced?

23. RACHAEL'S JUICE

Rachael is mixing some fruit juice for tomorrow's party. She has 46 quarts of 29% juice and a whole bunch of 17% juice. She thinks that 29% juice is too strong and 17% juice is too weak. She wants to make 25% juice. So she plans on mixing all 46 quarts of 29% juice with some amount of 17% juice in order to make 25% juice. How much 17% juice should she use?

24. TALIA'S TRULY TERRIFIC FRUIT PUNCH

Talia is mixing some punch for tomorrow's party, with a little help from her friends. Josalyn brought 24 quarts of a 16% punch, and Tanner brought 40 quarts of a 35% punch. Talia plans to mix these punches and add water. She wants her final punch to be 25% punch. How much water should she add?

Problem Set B, Version 2

1. THE DRAMA PRODUCTION

Daffany and her friends are in a drama production. At the beginning of the show, she and the others all line up behind Augie, march on stage, and then peel off from a marching circle until one person is left. Actually, every other person starting with the lead person peels off. The others keep marching around in the circle until it is their turn to peel off. As they peel off from the marching circle, they form a line from left to right across the stage. The final order is supposed to be Augie, Beatrice, Corrinne, Daffany, Estefan, Fiona, Gwendolyn, Heather, Ilana, Jackie, Kent, and Lorena. In what order do they need to line up originally?

2. SELLING STEAK

Karl ran a dairy delivery service with a lot of regular stops. He was a friendly, clever guy who liked to make up puzzles for his customers. One day he bought some steaks at one of his stops and figured he could make a small profit at a later stop. Freda was interested, so he told her, "I'm going to offer these boxes of steaks to you on sale. You need to figure out two things, my sale price per box and how many boxes there are." Karl continued, "If you multiply my sale price per box times the number of boxes, that's your total price. But remember, these boxes are on sale. The regular price per box is the same as the number of boxes. Your total price will be 75 times the amount you are saving on each box. When you add the regular price for each box to my sale price for each box, the result is 7 times the amount you are saving on each box. Now, I'm going to start unloading your order. Let me know by the time I'm ready to go if you want the steaks." How many boxes are there, and what is Karl's sale price for each box?

3. DICEY PRIMES

Of three dice, two are labeled normally: 1, 2, 3, 4, 5, and 6. The third die has sides labeled 2, 4, 6, 8, 10, and 12. Two dice are rolled and their values added. Find the ratio of the frequency of primes rolled using the first two (ordinary) dice to the frequency of primes rolled using the unusual die and one ordinary die.

4. ICE CREAM COUPONS

A local ice creamery gave the principal of the high school 295 ice cream certificates, good for a huge order of ice cream, to be awarded to any student with a B average or better. As it happened, 297 students had at least a B average. Since there were 2 more students than ice cream certificates, the principal started a long, agonizing process of giving them out and determining who didn't get one. Each day the principal chose 3 students at random. If all 3 were freshmen, they were each awarded a certificate. If 2 of the 3 were sophomores, the principal awarded the certificates to the 2 sophomores and put the third's name back in the hat. If all 3 were from a different class, he awarded all 3. If any other combination of students was selected, no certificates were awarded that day and the names went back in the hat. At the end of many days, there were only two B-average students left who had not received a certificate. At the beginning there had been 101 freshmen, 99 sophomores, and 97 juniors. From which class were the two remaining students?

5. HALFWAY LINES

In the grid below, exchange the places of one pair of numbers so that you can draw a vertical line that divides the grid into two parts with equal sums and a horizontal line that also divides the grid into two parts with equal sums.

1	3	2	4
5	7	6	8
9	11	10	12

Problem Set B, Version 3

1. LATE TO THE GATE

Tiffany hated being late for her plane, but her first flight into Atlanta's Hartsfield Airport was late. She went to the ticket counter to find out what gate her connecting flight left from. The ticket agent told her to go to gate 47, which was pretty far away. After walking a while she reached a moving sidewalk. She got on the moving sidewalk and continued walking. She looked at her watch and noticed it took her 16 minutes to get from the ticket counter to gate 47. Unfortunately, that wasn't fast enough; by the time she reached the gate, her plane had already left and was taxiing on the runway. She walked all the way back (without using the moving sidewalk) to the ticket counter to register a complaint and book a new flight. It took her 25 minutes to walk from the gate back to the ticket counter. About 2 hours later, she left the snack stand by the ticket counter and walked and rode the moving sidewalk back to gate 47. This time she didn't walk while she was on the moving sidewalk, and the trip took her 20 minutes. She knows she walks about 3 miles per hour. What is the speed of the moving sidewalk?

2. CARD ARRANGEMENT

You have ten cards, numbered 1, 2, 3, 4, 5, 6, 7, 8, 9, and 10. Your task is to arrange them in a particular order and put them in a stack, hold the stack in your hand, and then do the following: Put the top card faceup on the table, put the next card on the bottom of the stack in your hand, put the next card on the table, put the next card on the bottom of the stack, and so on, continuing to alternate cards that go on the table and under the stack, until all ten cards are on the table. That, of course, would be really easy to do, so the trick is, when the cards are put on the table they are to be in numerical order. In other words, the first card you put on the table will be number 1, the next card you put on the table will be number 2, the next card you put on the table will be number 3, and so on, until the last card placed on the table is number 10. In what order should the cards be arranged in the original stack so that this will happen?

Problem Set B, Version 3 *(continued)*

3. RECTANGLE RATIOS

A certain rectangle has its area, its perimeter, and the difference of its dimensions in a ratio of 15 to 12 to 4. Find the length and width.

4. TRIOMINOES

How many different figures can be made using six same-sized equilateral triangles placed so that at least one edge of each triangle adjoins the edge of another triangle?

See the examples below. Reflections and rotations don't count as different.

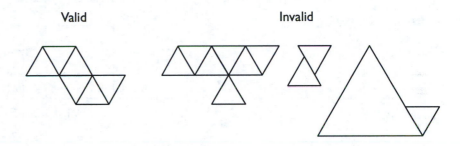

Valid   Invalid

5. NEMATOAD

No one seems to know how this game got its name, but it doesn't really matter. It is played like this: Two players place 25 beans between them. They then take turns picking up 1 to 4 beans until there are none left. Whoever picks up the last bean is the winner. You are going to go second. What should your strategy be?

Problem Set B, Version 4

1. **AMAZING NUMBERS**

From each square in this maze, you can travel in a straight line in one direction (vertically, horizontally, or diagonally) exactly the number of spaces indicated by the number in the square. For example, from the 6 in the lower left-hand corner, you could move vertically 6 spaces up to the 4, or 6 spaces diagonally to the 3, or 6 spaces horizontally to the right to the 2. Any of those would be considered one move.

Your task is to start at the 3 in the upper left-hand corner and reach the E in the lower right-hand corner in exactly six moves.

3	5	6	4	2	6	7	5
4	7	3	5	2	6	3	7
5	2	2	3	2	1	7	4
7	1	4	5	4	2	5	5
4	1	4	3	5	4	4	6
3	2	6	5	3	6	3	5
7	1	2	4	6	2	3	2
6	4	5	3	7	1	2	E

2. **LID ASTRAY**

In my kitchen cabinet I have lots of different-sized plastic containers, all unlabeled and all the same color. The lids for all of these (as well as for some containers that I don't even have anymore) are in a drawer. The other day, I needed several containers for freezing spaghetti sauce. I pulled out three different-sized round containers. I then went to the lid drawer and was confronted with a large mass of lids. I pulled out five lids, all different sizes. I knew that the lids for the three containers that I had were among the five lids. Devise a strategy for getting the right lids on the right containers as efficiently as possible. Outline your whole stategy, indicating what you will do next when a lid *fits* a container and what you will do next when a lid *doesn't fit* a container.

3. **RADIATOR**

The capacity of my car radiator is 24 quarts. The mixture of antifreeze and water in the radiator is 25% antifreeze. Since winter is approaching, I want a larger percentage of antifreeze. My mechanic suggests a 38% mixture. I therefore need to drain off some of the current mixture and replace it with pure antifreeze. How much of the current mixture do I need to drain off? (Answer exactly.)

Problem Set B, Version 4 *(continued)*

4. **LSAT TEST**

Problems similar to this one appear on the LSAT test, which is the test you must pass to get into law school.

A flagpole has spaces for seven colored flags arranged in a vertical line. Two of the flags are yellow, two are green, one is red, one is orange, and one is brown. Flags are to be placed on the pole under the following conditions: The orange flag is to be placed immediately below the brown flag. The red flag is not allowed to be immediately above or below either green flag. The two green flags must be together. The two yellow flags must not be together. The red flag is not allowed to be at the top or the bottom of the pole. How many acceptable flag arrangements are there?

5. **LARGE POWER OF TWO**

Perhaps you have seen the notation 4! (read "four factorial"), which means $4 \times 3 \times 2 \times 1$. Similarly, $10! = 10 \times 9 \times 8 \times 7 \times 6 \times 5 \times 4 \times 3 \times 2 \times 1$.

A factorial always starts with an integer and multiplies it by all preceding positive integers all the way down to one.

$$n! = n(n - 1)(n - 2)(n - 3) \ldots (3)(2)(1)$$

As you can imagine, 800! is a very large number. Find the largest power of 2 that divides 800! without a remainder. (*Hint:* This problem is easier to solve if you do not use a calculator. Figure out where the 2's come from.)

14 Evaluate Finite Differences

Finite differences are a means for determining the equation of a polynomial function by examining the changes in the dependent variable. This method does not work for trigonometric or exponential functions. You can analyze a polynomial function by considering the differences between successive values of the function. In linear functions, the successive values have a constant difference. In quadratic functions, the first layer of differences creates a linear function, for which the next layer of differences reveals a constant.

Note that we use the letters g, f, and e when doing second-degree equations. This is to avoid confusion with linear equations in which m and b are used and with the quadratic formula where a, b, and c are used.

When you solve a problem using finite differences, you will want to set up a master chart of differences using only variables. Compare the master to the specific differences for the function you are examining. This comparison will give you a choice of a number of equations to work with.

Finite differences should also be helpful in teaching students about the concept of variables and about functions. You will also notice a correspondence between finite differences and solving systems of equations.

You can apply finite differences to a number of functions in which the input and output are known but the function's equation is not. For example, finite differences can be applied to the price of chicken nuggets. If there's a constant difference in price corresponding with successive different-sized orders, you can calculate a price per nugget (the slope of the linear function). The number added to this (the y-intercept) might be the markup. Realize, of course, that this is a gross oversimplification. But polynomial functions are at least a starting point for many pricing strategies.

Good Quiz Problems

Any of the problems in Problem Set A, Version 2, would work fine. Give at least ten minutes per function if you are going to give more than one problem.

Text Problems

FAZAL'S FUNCTIONS

Fazal was creating input-output charts. He created these two functions, and he wanted to know the missing values. Can you help?

1.

x	y
0	3
1	7
2	11
3	15
4	19
5	−?−
137	−?−

2.

x	y
0	−4
1	1
2	12
3	29
4	52
5	−?−
137	−?−

LINEAR FUNCTION GENERAL CHART

Consider the general equation for a linear function, $y = fx + e$. Make a function chart for this equation, using the values 0 through 5 for x. Do this before reading on.

QUADRATIC FUNCTION GENERAL CHART

The general form for a quadratic function is $y = gx^2 + fx + e$. As you did for the earlier Linear Function General Chart problem, make the general function chart for this equation by substituting x-values from 0 through 5. (Note that when $g = 0$, this chart is the same as the one for $y = fx + e$.) Do this problem before reading on.

FAZAL'S SECOND FUNCTION

Find the equation for Fazal's second function on text page 354. Compare it with the general chart to find the values for g, f, and e. Then find the y-value for $x = 137$.

Text Problems *(continued)*

■ FIND THE FUNCTION

Find the equation for these two functions. (The first is Fazal's first function.) Then find the value of y when x is 5 and when x is 137. Do this before reading on.

1.

x	y
0	3
1	7
2	11
3	15
4	19
5	–?–
137	–?–

2.

x	y
0	–2
1	5
2	12
3	19
4	26
5	–?–
137	–?–

■ MORE FUNCTIONS

Find the equation for each function displayed in the charts below. Then use your equations to find the missing y-values, indicated by –?–. Work these before going on.

x	y
0	–5
1	–2
2	5
3	16
4	31
5	50
48	–?–

x	y
2	14
3	11
4	8
5	5
6	2
7	–1
82	–?–

Text Problems (continued)

HANDSHAKES

Suppose that at the first meeting of the House of Representatives, all 435 members shook hands with each of the other members. How many handshakes took place? Work this problem before continuing.

HOW MANY SQUARES?

Find a formula for the number of squares on any *n*-by-*n* checkerboard. Set up a pattern for this problem before continuing.

CUBIC FUNCTION GENERAL CHART

Generate the chart for the general form of a cubic function, $y = hx^3 + gx^2 + fx + e$. Do this before continuing.

Problem Set A, Version 2

1. **A COVEY OF FUNCTIONS**

Find the equation for each function.

a.

x	y
0	3
1	7
2	11
3	15
4	19
5	23

b.

x	y
0	4
1	−1
2	−6
3	−11
4	−16
5	−21

c.

x	y
0	3
1	5
2	15
3	33
4	59
5	93

d.

x	y
0	7
1	7.5
2	8
3	8.5
4	9
5	9.5

e.

x	y
0	2
1	0
2	−2
3	−4
4	−6
5	−8

f.

x	y
0	7
1	12
2	21
3	34
4	51
5	72

g.

x	y
0	−1
1	11
2	41
3	101
4	203
5	359

h.

x	y
0	−2
1	−3.5
2	−5
3	−6.5
4	−8
5	−9.5

Problem Set A, Version 2 *(continued)*

2. **BLOCK PYRAMID**

The drawings below show a 1-layer pyramid, a 2-layer pyramid, and a 3-layer pyramid made out of building blocks. "Visible" blocks are those that you could see if you were to walk around the pyramid. Notice that the first one has 1 visible block, the second one has 5 visible blocks, and the third one has 13 visible blocks. One hidden block is in the middle of the bottom layer of the third pyramid. How many blocks would be visible on a 35-layer pyramid?

 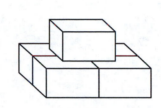 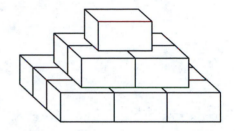

3. **A BEVY OF FUNCTIONS**

Find the equation for each function.

a.

x	y
0	1
1	3
2	5
3	7
4	9
5	11

b.

x	y
0	0
1	3
2	4
3	3
4	0
5	−5

c.

x	y
0	−7
1	−3
2	1
3	5
4	9
5	13

d.

x	y
0	−3
1	2
2	9
3	18
4	29
5	42

e.

x	y
0	1
1	−2
2	−5
3	−8
4	−11
5	−14

f.

x	y
0	−2
1	−7.5
2	−12
3	−15.5
4	−18
5	−19.5

Problem Set A, Version 2 *(continued)*

4. **HEXAGONAL NUMBERS**

The figures below represent the first four hexagonal numbers. Find a formula for the hexagonal numbers.

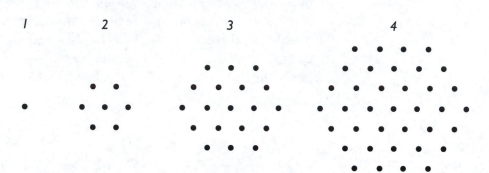

5. **A GAGGLE OF CUBIC FUNCTIONS**

Find the equation for each function.

a.

x	y
0	23
1	30
2	31
3	14
4	−33
5	−122

b.

x	y
0	8
1	11
2	16
3	17
4	8
5	−17

c.

x	y
0	−5
1	−3
2	13
3	49
4	111
5	205

d.

x	y
0	5
1	−2.5
2	−19
3	−41.5
4	−67
5	−92.5

Problem Set B, Version 2

1. WORMS

Wade "wuvs" worms. Every week, when exposed to sunlight for one hour, each worm splits into four parts. Each part becomes a full-fledged worm. Wade decides to run a worm farm. He figures he can sell his worms to bait and tackle shops in the area. Wade starts with 14 worms, and every week he exposes them to sunlight for one hour and watches them split. After six weeks, he takes all of his worms and packs them up. Each worm weighs 3 ounces. He packs them in 2-pound cartons and packs the cartons into cases: 40 cartons to a case. How many cases of worms does he have at the end of six weeks?

2. PRODUCE

At the grocery store, you can buy

1. An orange and an apple for the price of a grapefruit

2. An orange for the price of an apple and a peach

3. Two grapefruits for the price of three peaches

One orange would cost the same as a certain number of a certain fruit. How many of which fruit?

3. ANTIFREEZE

I like to keep my car's radiator full of an antifreeze-water mixture that is 25% antifreeze and 75% water. Recently, I left the cap off while I was driving, and a lot of the mixture leaked out. I filled up the radiator with 9 quarts of water (I was out of antifreeze) and then found that the mixture was only 12% antifreeze. What is the capacity of the radiator?

4. STRANGE NUMBER

There is a five-digit number that has the following strange property: Put a 2 in front of this five-digit number. Multiply the resulting six-digit number by 3. The product is a six-digit number made up of the original five-digit number followed by 2. What is the five-digit number?

5. TWO SEQUENCES

Consider the two sequences 21, 32, 45, 60, 77, 96, . . . and 1505, 1540, 1575, 1610, 1645, 1680, ... How many terms will it take for the first sequence to overtake the second sequence?

©2004 Key College Publishing, Instructor Resources: Crossing the River with Dogs, Johnson/Herr/Kysh

Problem Set B, Version 3

1. OLD CHEVY

My old Chevy Malibu loves super premium gas. Unfortunately, my sister drove it one day and topped it off with 5 gallons of regular gas. The tank holds 16 gallons. My car gets 17 miles per gallon. How many miles will I have to drive the car so that when I fill up the gas tank with super premium gas, the mixture in the gas tank will be at least 95% super premium gas?

2. COUNTY FAIR

John and Marcia took her two kids to the county fair. They first spent ⅙ of their money on adult admission tickets, and then $6.50 on the two child's tickets. They spent ⅜ of their remaining money on ride tickets. Then they spent $4.00 each on lunch. They spent ⅕ of their remaining money on a T-shirt for John. Marcia wanted a T-shirt too, so they bought another one for the same price. Then one kid found a quarter and the other one found a dime. They each gave the money to their mom, whereupon John and Marcia each gave each child a dollar. John and Marcia then spent ⅖ of their remaining money on ice cream sundaes. When they left the fair, they had to pay their parking tab, which amounted to $1.50 an hour. They had been there for seven hours. When they got to Marcia's house, John and Marcia had $3.15 left between them. How much money did they spend at the fair?

3. HOW MANY SEGMENTS?

There are n points on a piece of paper, no three of which are in a straight line. How many line segments can be drawn joining every possible point on the paper?

Problem Set B, Version 3 (continued)

4. **MYSTERY SUM**

The sum below features consecutive numbers in each column. The numbers can be consecutive reading down or reading up, but the entire column (including the digit in the answer) is consecutive in one or the other order. (*Note:* 0 and 9 are not considered to be consecutive.) Fill in the spaces. Two 5's have been filled in to help you.

$$
\begin{array}{r}
\underline{} \\
\underline{}\ \underline{}\ \underline{} \\
\underline{}\ \underline{}\ \underline{}\ \underline{} \\
\underline{}\ \underline{}\ \underline{}\ \underline{}\ \underline{} \\
\underline{}\ \underline{}\ \underline{}\ \underline{}\ \underline{}\ \underline{} \\
\underline{}\ 5\ \underline{}\ \underline{}\ \underline{}\ 5\ \underline{}\ \underline{} \\
+\quad \underline{}\ \underline{}\ \underline{}\ \underline{}\ \underline{}\ \underline{}\ \underline{}\ \underline{} \\
\hline
\underline{}\ \underline{}\ \underline{}\ \underline{}\ \underline{}\ \underline{}\ \underline{}\ \underline{}
\end{array}
$$

5. **HOLLYWOOD SQUARES**

Nine ordinary people get to be the celebrities on the game show *Hollywood Squares*. They sit in a tic-tac-toe board in three rows, with three people in the top row, three people in the middle row, and three people in the bottom row. The following clues refer to the positions of the nine people. Names beginning with the letters A through I are first names, and names beginning with J through W (not all letters are used) are last names. (*Note:* If a clue states that a person is between two other people, you don't know who is on the left and who is on the right, but you would of course know who is in the middle.)

1. Betty is between Madison and Kennedy.

2. Lincoln is between Anh and Ellen.

3. Jefferson is between Gregorio and Taft.

4. Frank is in the bottom row next to Roosevelt.

5. Isabel is under Gregorio and over Pierce.

6. Della is in the top row next to Howard.

7. Clara is in the middle row next to Nixon.

8. Isabel, whose name is not Madison, is on the far left side.

9. Anh is neither Pierce nor Wilson.

10. Della is not in the same column as Roosevelt.

Determine each person's full name and his or her seating position.

©2004 Key College Publishing, *Instructor Resources: Crossing the River with Dogs*, Johnson/Herr/Kysh

Problem Set B, Version 4

1. ANTS

A long stick is leaning against a wall. A group of ants spies the stick, and they figure they could climb up the stick to reach a hole in the wall. Unfortunately, the stick does not quite reach the hole. The ants measure the distance from the bottom of the stick to the bottom of the wall and find it to be 15 inches. They estimate that the top of the stick is 4 inches below the hole. So they decide to push the bottom of the stick toward the wall. They push the bottom of the stick toward the wall at a rate of ½ inch per minute. This causes the top of the stick to slide up the wall. It reaches the hole in 16 minutes. How long is the stick?

2. SEQUENCE RACING

Consider the two sequences 4, 8, 14, 22, 32, . . . and 5000, 5200, 5400, 5600, 5800, 6000, ... How many terms will it take for the first sequence to overtake the second sequence?

3. CORNER CAFÉ

The new waiter at the Corner Café had a problem getting orders straight. When he brought out the order to table 14 (a square table with one person on each side) every order was messed up. He brought french toast and strawberries to Karen, eggs and hash browns to Gil, an omelette and a muffin to Donna, and pancakes and sausage to Stan. No person received either item that they ordered, but no one received both of the items that were ordered by someone else. The four diners managed to sort out the confusion and rearrange the food. When order was restored, Karen was sitting between hash browns and pancakes, Donna was sitting between eggs and french toast, the person sitting on Stan's right had sausage, and the person across from Gil had a muffin. No one had ordered two main items (which are french toast, eggs, omelette, and pancakes). Who ordered what?

4. THE GREAT NUMBERINI'S NUMBER

"I am the great 'Numberini.' I am thinking of a five-digit number with five different digits. Place a 2 in front of my number to make a six-digit number. Multiply the six-digit number by 16. The answer is a seven-digit number that begins with my five-digit number, followed by a 2 and then a 0. What is my five-digit number?"

TEACHING RESOURCES

©2004 Key College Publishing, *Instructor Resources: Crossing the River with Dogs*, Johnson/Herr/Kysh

Problem Set B, Version 4 *(continued)*

5. SKIING ARIZONA

Last winter we went skiing in Flagstaff, Arizona. In the parking lot I noticed license plates from Arizona as well as from various other states. There were also some license plates from other countries. Some of the vehicles were trucks, and the rest were not trucks. Some of the vehicles had four-wheel drive (FWD), and the rest did not. Here are some other observations:

1. There were 56 total vehicles.

2. Every truck had FWD.

3. Thirty-five of the vehicles had FWD and 21 did not.

4. There were 20 trucks.

5. There were 42 vehicles from the United States, 28 of which were from Arizona.

6. There were four times as many Arizona trucks as trucks from other states in the United States.

7. There were 2 more non-FWD vehicles from Arizona than there were FWD vehicles that were not trucks and not from the United States.

8. The number of trucks was 8 less than the number of vehicles from Arizona.

9. There were 7 more Arizona trucks than there were trucks from other countries.

10. The Arizona vehicles that were not trucks were split evenly between FWD and non-FWD.

How many of the U.S. vehicles that weren't from Arizona were not trucks? How many foreign FWD vehicles were there?

15 Organize Information in More Ways

Much of problem solving involves organizing a problem's information in some different manner than it is presented in. For example, if you draw a diagram to solve a problem, part of what you do is to organize the information in a different way. Just as a picture is worth a thousand words, organized information presents a better picture. We are accustomed to much information being presented in linear form: Sentences, paragraphs, chapters, and books are all linear. Music is linear. On the other hand, some of the problems in this chapter are solved by organizing the information in two dimensions. The Mathlete example and the Three Squares problem in the student text are examples of two-dimensional organization schemes. Such schemes are easily represented on a piece of paper (as paper—at least the part useful to us—is two-dimensional).

At the end of the chapter we present tree diagrams (also presented in Chapter 2: Make a Systematic List). Tree diagrams are another good way to use two dimensions to organize a problem. Time is often one of the dimensions represented in a tree diagram. In the tournament example in the student text, the horizontal dimension represents time (successive rounds of the tournament) while the vertical placement signifies the person's placement within the play-off schedule. Tree diagrams are a wonderful way to solve problems that come up in probability and statistics. The student book includes problems involving drug testing, games of chance, coin flipping, and many other situations involving conditional probability.

Organizing Information is one of the broad "umbrella" themes of problem-solving strategies. Some of the strategies based on organizing information include systematic lists, eliminating possibilities, matrix logic, looking for a pattern, guess-and-check, unit analysis, algebra, and finite differences. Elements of organizing information are also present in working backwards. Organizing information makes a problem more manageable.

Reorganizing information might be a better name for this strategy. Information in a problem may already be organized in some way, but what you need to do is to approach the problem aggressively with the assumption that it can be organized better.

TV schedules are an example of how information was organized in a certain way before it occurred to someone to organize it differently.

TEACHING RESOURCES

©2004 Key College Publishing, *Instructor Resources: Crossing the River with Dogs*, Johnson/Herr/Kysh

For years, schedules were invariably written out as a linearly organized listing based on times:

8:00

(3) **Lifestyles of the Guess-and-Checkers:** TV's cleverest characters are highlighted in this display of designer jeans on Rodeo Drive. We also pay a visit to an ex-presidential dog.

(10) **Wheel of Diagrams:** So popular has buying a vowel become that a shortage arises. Imported *o*'s with umlauts (*ö*) make their debut.

(13) **All My Physical Representations:** Bodybuilder Storm is showing off his physical representations . . . yet again.

9:00

(10) **Crossing the River with Dogs (adventure):** The president stops off in Rocklin to seek wisdom from our heroes, Ted and Ken.

(13) **I Dream of Systematic Lists:** Art, Brad, Cindy, and Donna find several ways to line up at a drinking fountain, and Penny Stimes pays a return visit.

9:30

(10) **As the Diagram Draws:** Leslie finds out what Rob and Kelsey have known all along and what Mom, Gus, and Rover suspected. Never mind, just watch the show.

10:00

(13) **Wide World of Subproblems:** New show features loading a truck in Chicago, planning a Rocky Mountain camping trip, diagnosing a malfunctioning car in Phoenix.

10:30

(3) **Wyatt ERP:** Famous lawman seeks to solve problems the easy way.

(10) **The Simplesons:** At a new time—Bart and his family go head-to-head solving simpler problems.

More recently, most TV schedules have been revamped to list times along the top and stations on the left. This two-dimensional model is more efficient for presenting the essential information.

	8:00	8:30	9:00	9:30	10:00	10:30	11:00
(3)	Lifestyles of the Guess-and-Checkers					Wyatt ERP	
(10)	Wheel of Diagrams		Dogs	As the Diagram Draws		The Simplesons	
(13)	All My Physical Representations		I Dream of Systematic Lists		Wide World of Subproblems		

The length of the show is represented by the length of the bar that bears its name. In the linear scheme, the length of a show was not presented. The only way to determine a show's length was to look through the list until the next time a show appeared on that channel.

Notes on Text Problems and Problem Set A

THE THREE SQUARES

This was originally solved by systematic lists and eliminating possibilities in Chapter 3: Eliminate Possibilities. Now a solution is presented using Organizing Information. It is important that students know and see problems solved in a variety of ways. This is a good example of organizing the information differently so that the organization highlights different elements of the problem. Notice also that seeking contradictions (indirect proof) shows up in this problem.

DRUG TESTING

You may want to share with your students that an incorrect positive drug test is not a random occurrence. Very often, people who do not use drugs test positive for drugs because they have something in their body chemistry that produces the positive test—and they'll continue to test positive every time they take the test as long as their body chemistry remains the same. For example, testing hair samples for drugs has recently come under fire for placing people of color at a disadvantage. Because of body chemistry, more people of color have false-positive test results than do people who are white.

TWO BILLS

Many students will miss the point of this problem as written. It may be helpful to suggest looking for the maximum age Bill Junior could be and the minimum age Bill Junior could be. The same is true for Dolores's Age in Problem Set A, Version 2.

Good Quiz Problems

We recommend no quiz on this chapter.

Text Problems

THE THREE SQUARES

Three cousins, Bob, Chris, and Phyllis, were sitting around watching football on TV. The game was so boring that they started talking about how old they were. Bob (the oldest) noticed that they were all between the ages of 11 and 30. Phyllis noticed that the sum of their ages was 70. Chris (the youngest) burst out, "If you write the square of each of our ages, all the digits from 1 to 9 will appear exactly once in the digits of the three squares." How old was each person? Do this problem before continuing. See if you can organize the information in the problem in a new way.

©2004 Key College Publishing, *Instructor Resources: Crossing the River with Dogs,* Johnson/Herr/Kysh

FOUR COINS

Liberty is going to flip four coins at once: a penny, a nickel, a dime, and a quarter. How many ways are there for the four coins to come up? Make a tree diagram of all possible results before continuing.

©2004 Key College Publishing, *Instructor Resources: Crossing the River with Dogs,* Johnson/Herr/Kysh

HEADS AND TAILS

After Liberty counted all the ways she could flip four coins, she wanted to know the probability of getting two heads and two tails when flipping four coins. What is that probability? Solve this problem before continuing.

©2004 Key College Publishing, *Instructor Resources: Crossing the River with Dogs,* Johnson/Herr/Kysh

RED AND WHITE

A bag contains one white ball and two red balls. A ball is drawn at random. If the ball is white, then it is put back into the bag along with an extra white ball. If the ball is red, then it is put back into the bag with two extra red balls. Then another ball is drawn. What is the probability that the second ball drawn is red? Solve this problem before continuing.

©2004 Key College Publishing, *Instructor Resources: Crossing the River with Dogs,* Johnson/Herr/Kysh

Text Problems *(continued)*

■ **DRUG TESTING**

A company that conducts individual drug tests to identify the existence of certain substances in a person's system claims that its drug test is 90% accurate. This is to say, given a group of people who use drugs, the test will correctly identify 90% of the group as drug users. Furthermore, given a group of people who don't use drugs, the test will correctly identify 90% of the group as non-users. Those numbers sound pretty good on the surface, but suppose someone has just taken this drug test and the results are positive. What is the probability that he actually uses drugs?

To solve this problem, we need to know what percentage of the general population uses drugs. Let's use the company's numbers: To test the validity of its drug test, the company tested a sample group of people, knowing that 5% of this group did in fact use drugs. Solve this problem with a tree diagram before continuing.

■ **CARNIVAL GAME**

You are at a carnival. A man is offering a game of chance. He will charge you $3 to play the game. The rules are as follows: You are to reach into a bag that contains three orange marbles and seven green marbles and draw out two marbles. If you draw two green marbles, the man will pay you $2. If you draw one of each color, the man will pay you $3. If you draw two orange marbles, the man will pay you $12. Should you play this game? Remember, it costs $3 to play. Draw a tree diagram to analyze this situation before continuing.

Problem Set A, Version 2[1]

I. THE CLASSIC HOMEWORK EXCUSE

Arthur claims that his puppy chewed up his homework and then stepped all over it with muddy paws (his parents' couch can prove it!). The homework assignment had been to prepare a totals chart for donuts and coffee. As much of the chart as could be read is reproduced below. As a result of the chart being "dogged," the typesetter may have misread some of these numbers. Determine the price of one donut and the price of one coffee, and fix any numbers that are inaccurate. (*Note:* When you re-create this matrix, do it assuming that as few of the numbers are mistyped as possible. Without this assumption, someone could make this problem very easy—assume that all numbers are mistyped, and set new prices for coffee and donuts.)

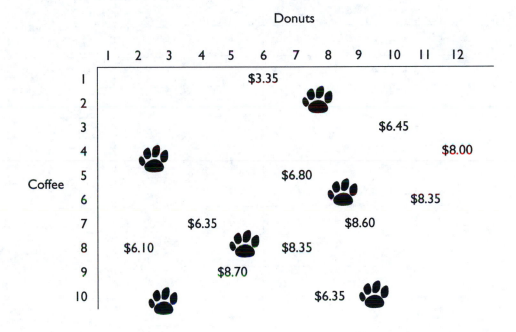

[1]We wish to thank Steve Weatherly, former teacher at Nevada Union High School in Grass Valley, California, and Allyson Angus Stewart, instructor at Santa Rosa Junior College in Santa Rosa, California, for writing Problems 7–13 in this problem set.

2. SON OF TWO-INPUT FUNCTIONS, PART I

The following are functions with two inputs. The first input is *x*, and the second input is *y*. What is the rule for calculating the output?

a.

Input	Output
5, 6	16
−5, 1	−9
8, 2	18
10, 1	21
4, 7	15
−5, −4	−14
−1, 6	4
−1, 0	−2
−1, 3	1
4, 8	16
8, 9	25
9, 8	26
2, 6	10
−3, 1	−5
−5, −2	−12

b.

Input	Output
−1, 3	10
4, 8	20
3, 6	15
8, 9	19
−3, 4	15
9, 8	15
2, 6	16
−3, 1	6
−5, −2	−1
10, 7	11
2, 3	7
5, 6	13
−4, 0	4
6, 5	9
1, 5	14

c.

Input	Output
−1, 3	4
4, 8	4
3, 6	3
8, 9	1
−3, 4	7
9, 8	−1
2, 6	4
−3, 1	4
−5, −2	3
10, 7	−3
2, 3	1
5, 6	1
−4, 0	4
1, 5	4
5, 5	−1

d.

Input	Output
10, 7	13
1, 5	−3
6, 5	7
2, 5	−1
6, 4	8
−3, 3	−9
3, 4	2
−4, −1	−7
7, 7	7
10, 0	20
9, 9	9
−5, −5	−5
−4, −4,	−4
−1, −1	−1
25, 4	46

TEACHING RESOURCES

©2004 Key College Publishing, *Instructor Resources: Crossing the River with Dogs,* Johnson/Herr/Kysh

Problem Set A, Version 2 *(continued)*

3. **SON OF TWO-INPUT FUNCTIONS, PART 2**

The following are functions with two inputs. The first input is x, and the second input is y. What is the rule for calculating the output?

a.

Input	Output
−2, 4	8
10, −1	17
8, 6	34
9, 8	42
3, 7	27
−5, −1	−13
−3, 4	6
−3, 0	−6
−4, 3	1
9, 2	24
−1, 6	16
2, 7	25
3, 6	24
3, 8	30

b.

Input	Output
1, 5	6
1, 8	9
0, 4	4
4, 5	21
5, −7	18
6, 9	45
7, 2	51
8, 9	73
3, 7	16
−1, 2	3
−4, 3	19
5, 5	30
−2, 3	7
7, 8	57

4. **DOLORES'S AGE**

Dolores told me several things about herself without telling me her age:

She said, "I played soccer in the 7-to-8-year-olds' league." I knew that she had to be 7 or 8 when the league started August 1 and that it continued through November.

She also said, "I was the best basketball player in the 8-to-9-year-olds' league." Basketball went from January through March, and you had to be 8 or 9 upon entering on January 1.

Later on she said, "I turned 8 in second grade." I knew she started second grade in September 1988, and she told me this in 1993.

How old could Dolores have been when this conversation took place?

©2004 Key College Publishing, *Instructor Resources: Crossing the River with Dogs*, Johnson/Herr/Kysh

5. LOG RIDE

Six people were on the log ride. There were three rows of seats and two people in each seat. The ages of the people on the left side differed in six-year increments, and the youngest person was in front. The person in the front right was twice as old as his seatmate. The person in the middle right was one and a half times as old as her seatmate. The person on the right back seat was half as old as her seatmate. The sum of the six people's ages is 145. How old are the youngest and oldest people?

6. COMPUTER PRINTOUT

The computer printout for the adults' and children's admission prices at the county fair is totally messed up, but still accurate if you can decode the information. The columns and rows are the prices for a certain number of adults and for a certain number of kids. What is the admission price for a child, and what is the admission price for an adult?

ADULTS

012345678910

C0061218243036424854854 60

H14101622283440465258 64

I2814202632384450566268

L31218243036424854606672

D41622283440465258647076

R52026323844505662687480

E62430364248546066727884

N72834404652586470768288

83238445056626874808692

93642485460667278849096

1040465258647076828894100

Problem Set A, Version 2 *(continued)*

7. FREEZER

There are five fudge pops and eight fruit pops in the freezer. If two pops are selected at random, find the probability that

    a. They are both fudge pops.

    b. One is a fudge pop and the other is a fruit pop.

    c. At least one is a fruit pop.

8. CHIPS

A bowl contains eight chips. Three of the chips are red, and the rest are blue. If two chips are drawn successively at random and without replacement, find the probability that

    a. The first chip drawn is red and the second chip drawn is blue.

    b. Both chips are red.

    c. One chip is red and the other is blue (the order in which they are drawn doesn't matter).

If the second chip drawn is blue, what is the probability that the first chip drawn was red?

9. BASKETBALL AND MATHLETES

The probability that Burbank's basketball team will beat Kennedy's is $3/5$. The probability that Burbank's Mathletes team will beat Kennedy's is $4/7$.

    a. Find the probability that the Mathletes team wins and the basketball team loses.

    b. If the basketball team wins, what is the probability that the Mathletes team will lose?

10. PAT PLAYS

At any given football game that Burbank plays, the probability that Patrick plays during the first half of the game is $5/6$. If he plays during the first half, there is only a $3/7$ chance that he'll play again during the second half. (He gets tired if he plays all out in the first half.) If he *doesn't* play during the first half, there is a $5/7$ chance that he'll play during the second half. Find the probability that Patrick played in the first half, given that he played in the second half.

©2004 Key College Publishing, *Instructor Resources: Crossing the River with Dogs*, Johnson/Herr/Kysh

Problem Set A, Version 2 *(continued)*

11. GLASSES

Suppose that the population of a certain town is made up of 45% men and 55% women. Of the men, 40% wear glasses, and of the women, 20% wear glasses. Given that a person chosen at random from this town wears glasses, what is the probability that the person is a woman?

12. TARGET SHOOTING

One of three men—Ed, Francis, or Grady—is chosen at random to shoot at a target. The probability that Ed will hit the target is $1/3$, that Francis will hit it is $1/4$, and that Grady will hit it is $1/2$. If you observe that the target is hit, what is the probability that the shot was made by Ed?

13. MYXEDEMA

Suppose that the incidence of myxedema (underactive thyroid gland) among people admitted to hospitals is 1 in 1500. And suppose that when a test used to determine whether a person has the disease is given to people who do have myxedema, it shows the presence of the disease in 90% of those tested (it yields a positive result). In the remaining 10% of those tested, the test yields a false-negative result. When the test is given to people *who do not have* myxedema, it shows the absence of the disease in 99% of those tested. In the remaining 1% of those tested, the test yields a false-positive result. If the test is used on a hospital patient chosen at random and the result is positive, what is the probability that the patient really has the disease?

14. STAMP COMBINATION

Suppose you have only an 8-cent stamp, a 10-cent stamp, a 7-cent stamp, and 3-cent stamp. What are the different amounts of postage you can make using these four stamps?

15. GOOD AND YUMMY RESTAURANT

The Good and Yummy Restaurant claims to make the best burritos in the world. Fortunately, most people know better. They start with a tortilla, put in beans, then give the customers some choices: jack cheese, cheddar cheese, or both kinds of cheese, and then hot sauce, barbecue sauce, or neither.

a. Set up a tree diagram showing all of the different types of burritos the Good and Yummy Restaurant offers.

b. Organize a chart to show the different types of burritos available.

Problem Set B, Version 2

1. ALONA'S BOARD FENCE

Alona told her son Faustin to paint the fence. Faustin whined and complained that painting whole boards bored him. So she said, "Okay, paint just $\frac{1}{2}$ of one board today. Tomorrow paint $\frac{1}{3}$ of a board and $\frac{2}{3}$ of a board. The next day you can paint $\frac{1}{4}$ of a board, and then $\frac{2}{4}$ of a board, and $\frac{3}{4}$ of a board. The next day do $\frac{1}{5}$ of a board, $\frac{2}{5}$ of a board, $\frac{3}{5}$ of a board, and $\frac{4}{5}$ of a board. The next day do all the sixths (less than 1) and then all the sevenths, and so on, until you have been painting for a total of 25 days." Altogether, how many boards did Alona expect bored Faustin to paint?

2. STOCKS

Four stockbrokers were talking about how their pet stocks were doing.

ALEX: My stock is worth 3 or 4 times as much as Billie Jo's stock.

BILLIE JO: Yeah, but my stock is worth \$3 or \$4 more than Cayla's stock.

CAYLA: But Alex's stock is worth \$3 or \$4 less than Derwood's stock.

DERWOOD: And my stock is worth 3 or 4 times as much as your stock, Cayla.

All the stock prices are whole dollar amounts. None of the prices contain a digit of 3 or 4. What is the price for each stock?

3. ZNORS

All Znors are Zmuds. Half of the Zorfs are Zmuds. There are 24 Zorfs. No Zlogs are Znors. There are 16 Zorfs that are not Zlogs. All Zlogs are Zmuds. Half of the Zlogs are Zorfs. No Znors are Zorfs. There are 9 Zmuds that are not Znors or Zorfs or Zlogs. There are 15 Znors. How many Zmuds are there? How many Zmuds are also Zlogs and Zorfs?

Problem Set B, Version 2 *(continued)*

4. **CONFERENCE**

At a company meeting of photocopy repair technicians, there were 17 people from four different cities (New York, Peoria, Topeka, and San Diego). The facts below refer to these 17 people.

1. More than half of the participants were from the Midwest.

2. The farther east the city, the more participants.

3. There was at least one person from each location.

One person was due to be transferred from one city to a city farther east, but all of the above would still be true. The person being transferred was given a going-away potluck by the other repair technicians in his office. Unfortunately, they all brought chips. From which city to which other city was the transfer?

5. **FOUR BY FOUR**

In a four-by-four grid, place the numbers from 1 to 16 with these constraints:

a. No row or column contains two consecutive numbers.

b. No row or column contains two multiples of 4.

c. There are no multiples of 3 in either row 1 or row 3.

d. The numbers 2 and 13 are in the third column.

e. The corners all contain prime numbers.

f. The numbers 6 and 11 are in the bottom row.

g. The numbers 4 and 15 are in the first column.

h. The numbers 14 and 10 are in the same row.

i. No two adjacent columns have the same number of odd numbers.

TEACHING RESOURCES

©2004 Key College Publishing, *Instructor Resources: Crossing the River with Dogs,* Johnson/Herr/Kysh

Problem Set B, Version 3: Dad's Birthday

1. DEAR OLD DAD

My father liked to puzzle people. He indeed was very puzzling. For example, he had a new puzzle cooked up for the birthday party we threw him. Attending the birthday party were my Aunt Jasmine, my Uncle Pedro, my grandma, and all of us kids. (Grandma is the mother of my dad, Aunt Jasmine, and Uncle Pedro.) My dad said, "Suppose you multiply the digits in my age. Subtract that from my age and the answer is 18." Basically, our response was "Big deal!" However, he continued, "I did the same thing when I was my brother's age, only the answer was 16." Again, our basic response was "Big deal!" Undaunted, my father continued speaking. "If I do it again when my brother is my age, the answer again will be 16." It started sounding intriguing now, because nobody had heard anything from his brother, and I sure didn't know what age Uncle Pedro was. Then my grandma spoke up. "It's weird, but I just did the same thing with my age and the answer is also 18. If I'd done it when I was my daughter Jasmine's age, the answer would have been 16. And if I do it again when Jasmine is my age, the answer will be 16 again."

How old are my Uncle Pedro and my Aunt Jasmine?

2. SIXTEEN CANDLES

We put 16 candles on Dad's birthday cake: 4 each of blue, green, yellow, and purple. Just for fun, we decided to arrange them in a line so that primary colors were only next to related secondary colors. What order were they in?

3. REAL DOLLS

My cousin Pat was also at the party. Pat brought a collection of 36 dolls. She gave us some clues about the dolls: "Some of them are Barbies; in fact, half of the female dolls are Barbies. Half of the Cabbage Patch dolls are female. There are as many females that are not Barbies or Cabbage Patch dolls as there are Cabbage Patch dolls. Half of the dolls are not female. All of the dolls except two (which I found at a garage sale and whose gender I found it impossible to determine) are either male or female. One of those two genderless dolls is a Cabbage Patch doll." How many male, non–Cabbage Patch dolls are there?

Problem Set B, Version 3: Dad's Birthday (continued)

4. WHH

Another cousin, Dave, was also at Dad's party. He said he found this problem on the bulletin board at work:

Find pairs of integer values that work for this equation.

$$wh + h + h = w^2 \qquad \text{(Note: } wh \text{ means } w \text{ times } h.\text{)}$$

5. ONE HUNDRED FACTORS

Dad was impressed at our problem-solving ability so far. He was reading a magazine after dinner and found this problem in the magazine for us to solve: What is the smallest number that has 100 different positive, integral factors, including 1 and itself?

©2004 Key College Publishing, *Instructor Resources: Crossing the River with Dogs,* Johnson/Herr/Kysh

Problem Set B, Version 4

1. MARBLES

I overheard this story on the playground one day. Three kids were talking. I heard the voices but didn't see who was talking. The voices were discussing a marble match that had just taken place.

One voice said, "Wow, what a weird game. First Zeus won ½ of Venus's marbles."

Another voice said, "Then Venus won ⅓ of David's marbles."

A third voice said, "Then David won ¼ of Zeus's marbles."

Then one of the three voices said, "We all ended up with the same number of marbles. I started with 34 marbles."

How many marbles did each kid end up with?

2. COUSINS

Five kids from the Gleane family (Tom, Bonnie, Janet, Stefan, and Michael) went to a family reunion with their cousins, the Ricannas (Ryan, Torrey, Will, Daniel, and Gary). They stayed in a townhouse near a lake for five days. There were five great hikes to take: around the lake, to the top of the waterfall, to the top of Lover's Leap, along the river, and to the store. Each day the cousins paired off, with one Gleane cousin matched up with one Ricanna cousin, and went for a hike. Arrange a daily hiking schedule for the five days so that each Ricanna is matched once with each Gleane and all ten kids get to go on each hike once.

3. ANT GRAPEVINE

Em, Mame, and Bea are all ants and live in an anthill. They learned a new verse to the song "The Ants Go Marching" and wanted to teach it to all of the other ants in the anthill. However, they didn't want to do *all* the teaching, so they devised a plan to divide the labor. Each of the three ants would teach the song to two other ants on the first day. These three ants would then stop teaching, and the six ants that learned the song on the first day would each teach two new ants on the second day. The plan was put into effect, and the teaching and learning continued for days and days. Each ant that learned the song on one day taught two new ants the next day. Each ant taught for only one day. After how many days would all four billion ants in the anthill know the song?

©2004 Key College Publishing, *Instructor Resources: Crossing the River with Dogs*, Johnson/Herr/Kysh

4. **LONG LIST OF NUMBERS**

The numbers from one to one billion were written out in alphabetical order. (The word "and" was not used, and no dashes were used, so the number 431 was written as four hundred thirty one, and 54,187 was written as fifty four thousand one hundred eighty seven.) What was the last number written in the list?

5. **ALPHABET SOUP**

The figure below shows the alphabet split into regions. The regions are formed by two circles, two rectangles, two squares, and two equilateral triangles. Vertices of the rectangles, squares, and triangles are noted by dots. Don't count any smaller triangles or quadrilaterals that are formed by the crossing of the lines. For the purposes of this problem, all references to rectangles refer to rectangles that are not squares.

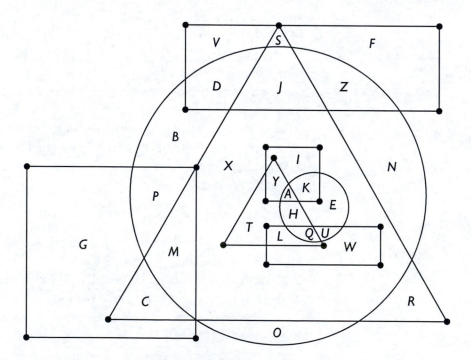

(This problem is continued on the next page.)

Problem Set B, Version 4 *(continued)*

The five clues below each describe a different letter. Take these five letters and rearrange them to form a word. (By the way, the word is not plural.) No two clues refer to the same letter.

1. A letter that is inside one less triangle than square and inside one more circle than triangle

2. A letter that is inside the same number of triangles and rectangles and inside the same number of squares and circles and is inside fewer squares than triangles

3. A letter that is inside more circles than triangles and inside more triangles than both squares and rectangles

4. A letter that is inside a total of four figures but only two different kinds

5. A letter that is inside more figures than any other letter and is not in a rectangle

©2004 Key College Publishing, *Instructor Resources: Crossing the River with Dogs,* Johnson/Herr/Kysh

16 Change Focus in More Ways

In his inaugural speech in 1961, President John Kennedy said, "Ask not what your country can do for you; ask what you can do for your country." His statement illustrates well the concept of changing focus. He wanted people to focus on themselves, not the government, as the solvers of the country's problems.

Changing Focus is one of the three major problem-solving themes presented in this book. This chapter is divided into three sections, each of which focuses on a different aspect of the Changing Focus theme: Change Your Point of View, Solve the Complementary Problem, and Change the Representation. *Changing your point of view* means looking at a problem from another perspective. For example, instead of focusing on the time spent driving, look at the distance driven. (Instead of asking what your country can do for you, ask what you can do for your country.) *Solving the complementary problem* means focusing on what *is not* there rather than focusing on what *is* there. For example, to evaluate the effectiveness of a medicine, look for instances when it is not effective. (Instead of asking what your country can do for you, ask what your country cannot do for you.) *Changing the representation* means finding another way to represent a problem's information so that you can solve the problem more efficiently or understand it better. Drawing a diagram and using manipulatives are two ways to change the representation and thus change focus.

Other strategies that reflect the Changing Focus theme are subproblems (looking at the components of the problem), easier related problems (looking at other problems first), and working backwards (looking at the problem in reverse).

Changing focus is a strategy we use often in our lives, not necessarily in concrete problem-solving situations, but in more subjective realms. A parent might have to change focus to deal with a child's misbehavior. Instead of focusing on the behavior—say, teasing—the parent may deal with the problem more effectively by focusing on the cause. Perhaps the child is jealous of the person he or she is teasing. Focusing on the child's feelings may help the child learn more mature ways of dealing with jealousy.

Robert Kennedy said, "Some people look at this country the way it is and say, 'Why?' I look at this country the way it can be and say, 'Why not?'" Besides being good rhetoric, his statement suggests that looking beyond perceived constraints in a problem is often a sign of genius and good leadership.

Note on a Text Problem

 **NINE DOTS**

In this problem, people often feel constrained to stay within the bounds created by the dots. Creative problem solvers look beyond perceived constraints. This principle is simply and effectively demonstrated by this problem, which can serve as a lead-in to dealing with the concept of constraints, real or imagined.

Good Quiz Problems

We recommend no quiz on this chapter.

Text Problems

 NINE DOTS

Without lifting your pencil from start to finish, draw four line segments through all nine points.

· · ·

· · ·

· · ·

The solution has nothing to do with how wide the dots are or with the idea that possibly the lines determined by them are not parallel. The points are mathematically defined—they have no width, and they determine sets of parallel lines.

Work this problem before continuing.

THE HUMAN FACTOR

Mayra is a human computer. She has appeared on talk shows to show off her amazing ability with numbers. One type of problem Mayra is very adept at solving is this: A person from the audience will give Mayra a number, and Mayra will immediately be able to tell how many one-digit **factors** that number has. For example, if you were in the audience and you said 50, Mayra would say 3, because 50 has 3 one-digit factors (namely 1, 2, and 5).

One day, Mayra was on a well-known talk show, and some wise guy in the audience asked Mayra to tell him how many one-digit factors the numbers from 1 to 100 had. The answer was not 9, because Mayra had to count each factor for a particular number, then add that count to the number of factors for each of the other numbers from 1 to 100. For instance, even though the factor 5 appears in the factorization of 50, it also appears in the factorization of 45, so it must be counted once for each of those numbers. Mayra quickly "programmed" her brain to give her the answer, and she had it in a few moments. What was her answer?

Work this problem before continuing.

Text Problems *(continued)*

AVERAGE SPEED

Jacques left his home in Austin and drove to San Antonio. On the way there, he drove 40 miles per hour (there was a lot of traffic). On the way back he drove 60 miles per hour. What was his average speed? Work this problem before continuing.

TERM PAPER

Seiko had to read five books for her 20th-Century American Lit. class. She then had to write a paper about the role of the family as presented in the books. She had narrowed down her choices to six books.

Of Mice and Men by John Steinbeck

The Bean Trees by Barbara Kingsolver

The Joy Luck Club by Amy Tan

The Color Purple by Alice Walker

A Lesson Before Dying by Ernest J. Gaines

Native American Testimony, edited by Peter Nabokov

In how many ways could Seiko choose the five books? Work this problem before continuing.

AREA

Find the area of the shaded region. Work this problem before continuing.

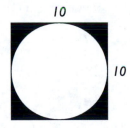

Text Problems *(continued)*

THE TENNIS TOURNAMENT

A big regional tennis tournament in New Orleans drew 378 entries. It was a single elimination tournament, in which a player was eliminated from the tournament when she lost a match. How many matches were played to determine the champion? Work this problem before continuing.

CARD ARRANGEMENT

You have ten cards numbered 1, 2, 3, 4, 5, 6, 7, 8, 9, and 10. Your task is to arrange them in a particular order and put them in a stack, hold the stack in your hand, and then do the following: Put the top card on the table faceup, put the next card on the bottom of the stack in your hand, put the next card on the table, put the next card on the bottom of the stack, and so on, continuing to alternate cards that go on the table and under the stack, until all ten cards are on the table.

That, of course, is easy to do. The trick is to lay the cards on the table in numerical order. In other words, the first card you put on the table will be number 1, the next card you put on the table will be number 2, the next card you put on the table will be number 3, and so on, until the last card placed on the table will be number 10. In what order should the cards be arranged in the original stack so that this will happen?

See if you can represent this problem in a completely different way (we used a diagram) and solve it more easily. Work this problem before continuing.

TEACHING RESOURCES

Problem Set A, Version 2

1. BASIC HEX

Without lifting your pencil from start to finish, draw 4 line segments through all 7 dots. The solution has nothing to do with how wide the dots are or the possibility that the lines determined by the dots are not parallel. The dots are mathematically defined—they have no width, and they determine sets of parallel lines.

2. TEN-DOT PUZZLE

Without lifting your pencil from start to finish, draw 5 line segments through all 10 dots. The solution has nothing to do with how wide the dots are or the possibility that the lines determined by the dots are not parallel. The dots are mathematically defined—they have no width, and they determine sets of parallel lines.

3. FOURTEEN-DOT PUZZLE

Without lifting your pencil from start to finish, draw 6 line segments through all 14 dots. The solution has nothing to do with how wide the dots are or the possibility that the lines determined by the dots are not parallel. The dots are mathematically defined—they have no width, and they determine sets of parallel lines.

4. HOURGLASS

Without lifting your pencil from start to finish, draw 4 line segments through all 8 dots. The solution has nothing to do with how wide the dots are or the possibility that the lines determined by the dots are not parallel. The dots are mathematically defined—they have no width, and they determine sets of parallel lines.

5. FRUIT STAND

The first customer bought $3.00 worth of oranges. She bought them in a bag, so she didn't know that the price came out to $0.40 per pound. The next customer also bought a $3.00 bag of oranges, but he paid $0.50 per pound for his oranges. Of the oranges sold so far, what has been the average price per pound?

Problem Set A, Version 2 *(continued)*

6. BADMINTON TOURNAMENT

Bobbi needs to select six badminton players to represent the school at a tournament. She can choose from Suvash, Debi, Amber, Mirn, Amando, Ravi, Brandt, and Kendra. In how many ways can she choose her six players?

7. DANCING HEARTS

Pauline has seven dancers to choose from to fill the positions as playing cards in a production based on *Alice in Wonderland*. The cards are all hearts, from two through eight. The dancers' names, in order from the shortest to the tallest dancer, are Brittney, Alyse, Katie, Aubrey, Wynnter, Jennifer, and Summer. How many different groups of dancers can Pauline set up as the cards?

8. AREAS OF SHADED REGIONS

Find the area of the shaded region. The cutouts on the ends are semicircles.

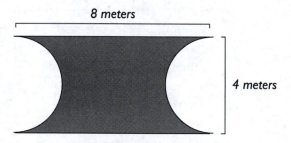

9. MORE AREA

Find the area of the shaded region.

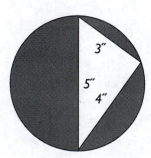

Determine whether the length of the diameter is 5 inches.

TEACHING RESOURCES

©2004 Key College Publishing, *Instructor Resources: Crossing the River with Dogs*, Johnson/Herr/Kysh

Problem Set A, Version 2 *(continued)*

10. ATTENDANCE FIGURES

The attendance figures for Durango High were off last week. Usually there is a 97% attendance rate. However, on Monday it was 91%, on Tuesday it was 94%, on Wednesday it was 93%, on Thursday the attendance was 95%, and on Friday it was 92%. How many student-days were missed last week? (The enrollment is 1800 students.)

11. LIGHT BEARINGS

Sitting in a box are 141 ball bearings. A quality-control worker claims that he accidentally dropped a defective ball (too light) into the box. Using a balance scale, find the minimum number of weighings it will take to find the defective ball.

12. HEAVY BEARING

You have 12 ball bearings, one of which is heavier than the others. What is the minimum number of weighings you must do on a balance scale before you can be certain of finding the heavier one?

13. PADUCAH HALF-TIME

The 12 cheerleaders of Paducah College do a maneuver at halftime. They run onto the field and run in a circle around the mascot. As the first person runs by, the mascot raises one paw. When the second person runs by, the mascot raises the second paw. The mascot then lowers both paws at once, pointing at the third cheerleader going by. As the mascot points, that cheerleader peels off and goes over to the 50-yard line, where the cheerleaders will start another routine. The cheerleaders continue to run around the mascot in a circle, and the mascot repeats its actions until all 12 cheerleaders have peeled off. The 12 are Araceli, Britt, Charissa, Darryl, Evan, Fran, Gabby, Henny, Izzy, Johanna, Kara, and Les. In what order should they line up off-field in order to peel off in the correct order (which happens to be in alphabetical order)?

©2004 Key College Publishing, *Instructor Resources: Crossing the River with Dogs*, Johnson/Herr/Kysh

14. WALKING HOME FROM SCHOOL

Every day Mom used to sit on the front porch and wait for us to walk the 3 miles home from school. The walk took us about 45 minutes. Our dog, Samson, waited for us at school every day. As soon as we walked out the school's doors, he would greet us and then run home. It only took him about 10 minutes to get home. As soon as he let Mom know we were out of school, he would come back and meet us on our way home, then turn around and run home to report to Mom on our progress. He continued this back-and-forth the whole time we walked home, even as we turned the corner at Mr. Jansen's house and came up the front walk. How far did Samson run each day?

15. MORE TOOTHPICKS

Using six toothpicks, make six triangles of the same size.

16. PERFECT CUBES

How many perfect cubes are there between 100 and 10,000,000?

17. SOME COMPLEMENTARY EVENTS

Find the complementary event for each situation.

a. A family has three kids, all girls.

b. The Family family watches three movies, all with "Friday the Thirteenth" in the title.

c. You find six coins, and none of them are dimes.

d. You meet four new people this week, and all are from Pennsylvania.

18. RUNNING GEORGE

George tries to run every day. He realizes that his body needs a break from running about twice a week, and usually his day-to-day schedule takes care of those breaks. He normally runs 6 miles each day. However, last week he couldn't run on Monday because he had to take Sue to the airport. On Wednesday, he cut his run short by about $1^1/_2$ miles because he twisted his ankle on an icy sidewalk. He then missed Thursday, but ran every day the rest of the week. He also cut his Saturday run short by 2 miles to make sure he could get Emily to her soccer game on time. How far did George actually run this week?

©2004 Key College Publishing, *Instructor Resources: Crossing the River with Dogs,* Johnson/Herr/Kysh

19. **IS SHE OR ISN'T SHE?**

Brand A early pregnancy test is accurate 92% of the time when it shows positive. Brand B is accurate 95% of the time when it shows positive, and accurate 87% of the time when it shows negative. If a woman tests positive with brand A and negative with brand B, what are the chances she really is pregnant?

20. **ALPHABET FLASH CARDS**

A deck of alphabet flash cards contains 26 cards. Try this challenge: Count off the cards. As you count the first two, place them on the bottom of the deck. The third one is placed faceup on the table. Continue doing this until you've placed all 26 cards on the table. In what order would the cards have to be arranged to be placed on the table in alphabetical order?

©2004 Key College Publishing, *Instructor Resources: Crossing the River with Dogs*, Johnson/Herr/Kysh

Problem Set B, Version 2:
The Treasure of Mount Nessum-Sar

1. RECTANGULAR TILES

The tomb of Queen Neiledam is reputed to hold the missing clues leading to the buried treasure of the mythical Mount Nessum-Sar. By using the tiles described below, form a rectangle—no overlaps, no gaps. Place this rectangle on the floor of the crypt. The scroll buried with the queen will reveal where to hold a torch, the light of which will reflect off the surface of the tiles onto the rough texture of the western wall of the crypt. The resulting design will be a map leading to the location of the buried treasure. The dimensions of the tiles are as follows:

9×11	5×9	2×8	5×6	4×7	2×9
11×13	4×5	4×8	5×12	4×7	2×9
7×13	2×5	3×9	1×20	4×7	

Form these small rectangles into a large rectangle, then go search for the treasure.

Unfortunately, after you solved the tile problem, the map was not revealed to you. Instead, a great genie appeared. He said, "I am the great Tenneb. If you can solve the next puzzle, Consecutive Integers, I will show you the map that leads to the buried treasure."

2. CONSECUTIVE INTEGERS

The genie said, "Tell me all of the integers between 1 and 600 that cannot be represented as the sum of two or more consecutive positive integers."

"I don't understand," you protested.

"All right," Tenneb replied. "I will give you an example: 3 can be represented as 2 + 1, and 12 can be represented as 3 + 4 + 5. But I want all the numbers that cannot be represented in this way."

Find all such numbers.

After you solved this puzzle, Tenneb turned to you and said, "Well, little one, you did very well. But I have another puzzle for you. It's called Alphabetical Numbers, and if you solve it, I will lead you to the treasure personally."

3. ALPHABETICAL NUMBERS

"Write out the numbers from 100 through 999 in alphabetical order. Don't use the word *and*. So, for example, write the number 431 as four hundred thirty-one. Next to your list, write the numbers from 1 through 900 in numerical order. Which number(s), if any, occupy the same position in both lists?"

©2004 Key College Publishing, *Instructor Resources: Crossing the River with Dogs,* Johnson/Herr/Kysh

Problem Set B, Version 2 *(continued)*

"But writing out all those numbers could take all night," you protested.

"Well then, little one, see if you can find the answer without writing out the whole list."

When you finished, the genie said, "Very good. Now I have just one more puzzle for you—the Say the Magic Word puzzle. If you solve this one, there will be a big surprise in store."

4. SAY THE MAGIC WORD

"I am thinking of five words," the genie said. "The five words are

PINT   RATE   TURN   NEST   WIND

"One of these words is the magic word. Your task is to figure it out."

"I need a hint," you said.

"All right," Tenneb said. "If I tell you any one letter in the magic word, then you can tell me the number of consonants in the magic word."

What was the magic word?

You solved this puzzle also and said, "Okay, I am ready for my big surprise. What is it?"

"Little one, you are in for a big surprise."

5. WHERE SHOULD YOU STAND?

You were shown into a great cavern. "Counting yourself, you will find 350 people in this cavern," the genie said.

You looked around and said, "These don't look like people, they look like mummies."

"Yes, they are mummies. Now, here's your next puzzle: You and the mummies will all line up in a single line. I will go down the line and count 1, 2, 3, 4, 5, 6, and so on. Every even-numbered person (or mummy) will step out of line. Whenever I reach the end of the line, I will go back to the beginning of the line and continue counting from wherever I left off. The even numbers will still step out of line. The last person left in line will get the buried treasure. You can stand anywhere you want, little one, but choose wisely."

The genie then shouted, "Line up!" as you worked furiously trying to figure out what position to be in. The mummies all began to shuffle into position. The genie turned to you and said, "Well, little one, where do you wish to stand?"

What position in line is the lucky winner?

Problem Set B, Version 3

1. LARGE CORPORATION

In a certain corporation, $2/3$ of the employees are men and $1/3$ are women. Of the men, $5/8$ are college educated and $3/8$ are not. Of the women, $2/5$ are college educated and $3/5$ are not. Of the college-educated employees, $5/6$ are in management. Of all non-college-educated employees, $1/9$ are in management. Of all managers, $1/3$ are women. If all of the college-educated women are in management, what percentage of the non-college-educated managers are women?

2. MOVIE STARS

Four actors are starring in a movie. Their first names are Bill, Ted, Grim, and Don. Their last names are Preston, Logan, Reaper, and Thanes. Their shirts are red, green, yellow, or blue. The patterns on the shirts are plain, striped, checkered, or print. Four kids who watched the movie made comments about the actors. Each comment referred to all four actors.

KID 1: I saw Bill, a person in a plain shirt, Logan, and a person in a green shirt.

KID 2: I saw Preston, a person wearing a blue shirt, another wearing a yellow shirt, and Ted.

KID 3: I saw Grim, a person wearing a checkered shirt, a person wearing a red shirt, and Don.

KID 4: I saw a person wearing a print shirt, a person wearing a red shirt, Grim, and Reaper.

Ted is not wearing the red shirt. Don is not wearing the plain shirt. Logan is not wearing the yellow shirt. Determine each person's full name and color and style of shirt.

3. SUMSUMS

The sumsum of a number is the sum of the number's digits added to the number itself. For example, the sumsum of 15 is $6 + 15 = 21$. How many of the numbers from 1 to 500 are sumsums of some other number?

TEACHING RESOURCES

©2004 Key College Publishing, *Instructor Resources: Crossing the River with Dogs*, Johnson/Herr/Kysh

Problem Set B, Version 3 *(continued)*

4. **WRESTLING NEWSLETTER**

Beth loves to watch wrestling on TV. She has a lot of friends who like wrestling too. Beth gets to watch it more than they do, though, so she often writes a short newsletter for them, detailing the results of recent wrestling matches. The other day she caught the tail end of a wrestling match in which a wrestler named Mango Mulch beat another wrestler named Gyurk. Gyurk was so sad that he cried. Beth wrote the article for the newsletter and used this headline:

MANGO MULCH BELTS GYURK: WHINE

After she wrote the headline, Beth noticed that her headline contained five 5-letter words. The weird thing was, there is another 5-letter word that shares exactly two letters with each word in the headline. What is that word?

5. **THREE-DIGIT TRIANGLES**

In the figure below, circles are connected to lines that form a whole bunch of triangles. Fill in the circles with the digits 1 through 9. When you are done, each triangle will form a three-digit number such that the difference between the smallest digit and the second-smallest digit is greater than 2, and the difference between the second-smallest digit and the largest digit is greater than 2. For example, one of your numbers could be 418, because the difference between 1 and 4 is 3 (more than 2) and the difference between 4 and 8 is 4 (also more than 2). On the other hand, 863 could not be one of the numbers. The difference between 3 and 6 is 3, which is fine, but the difference between 6 and 8 is only 2. The digit 3 has been placed in the figure to get you started.

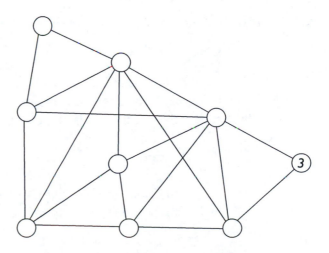

Problem Solving Strategies: Instructor's Resource Book and Answer Key, 2nd Ed. © 2000 Key Curriculum Press

Problem Set B, Version 4

1. REUNION

Five people attended a high school class reunion. To get there, one person traveled for one hour, another person traveled for one and a half hours, another for two hours, another for two and a half hours, and the last for three hours. Three of the people came in the same car (with one person driving the car and picking up the other two people at different places along the way). The other two people came in a plane (with one person flying the plane and picking up the other person along the way). The car and the plane each traveled at a constant speed, and the plane's speed was five times the car speed. Each speed was a whole number of miles per hour. The five people traveled a total of 1342 miles. What was the speed of the car? (*Note:* Travel times do not include stops and starts, takeoffs and landings, or acceleration and deceleration.)

2. POKER

Five poker hands were dealt, each containing five cards.

The first hand had a full house, which included two kings. This hand had the six of clubs, which was the only black card in the five hands.

The second hand had a diamond straight flush.

The third hand had a straight. This hand could have been a straight flush that would have beaten the second hand, but the highest card was the wrong suit.

The fourth hand had two pair. All five cards were of equal or lower denomination than any card in any other hand and lower than the two unused red cards.

The fifth hand had a heart flush.

Determine the cards in each of the five hands.

Note: A full house features two of a kind and three of a kind. A straight is five cards in consecutive order, but not all five cards are in the same suit. A flush is five cards of the same suit. A straight flush is five cards of the same suit in consecutive order. Two pair is two different sets of two of a kind and a fifth card that does not match either pair. The highest-ranking card is the ace, followed by the king, queen, jack, ten, nine, and so on, down to two. An ace can be used as a low card or a high card in a straight (such as A, 2, 3, 4, 5, or 10, J, Q, K, A).

Problem Set B, Version 4 *(continued)*

3. **BAND**

The director of a marching band is trying to decide what sort of rectangular arrays her band can form. She is interested in rectangles that have the same number of band members standing on the outside edge as the number of band members standing in the interior. What are the possible dimensions of rectangular arrays that allow this?

4. **DICE PRODUCTS**

Kathlan is a third grader in Ms. Latimer's class. Yesterday Ms. Latimer gave the class an assignment to practice their addition and multiplication. Each student was given a pair of dice. The students were to roll the dice and multiply the two numbers. They were to do this seven times and then add the seven products. Kathlan did this and got a total of 39. She noticed that none of her products were repeated. What is the probability that her list of products included an 8, a 15, or both?

5. **BOOK CHOICE**

Max works in the library. Every day before he goes home, he picks a book to read at home. (He reads a book every day.) This week he is interested in science fiction. The science fiction books are arranged in alphabetical order by the authors' last names. It turns out that there are 13 books with authors' names for each letter of the alphabet: There are 13 books with author names starting with A, 13 with B, and so on. Max has a strange way of picking out the book he will read. Every day he does it differently. Today he uses this system: He walks down the row of books and takes out every second book and throws it on the floor. When he gets to the end of the row, he goes back to the beginning. If he throws out the last book in the row, then he skips the first book when he starts again at the beginning. If he skips the last book, then he throws out the first book and continues, throwing every second book on the floor. He continues in this way until there is only one book left on the shelf. That is the book he takes home to read (after picking up all the books and arranging them on the shelf again). What is the first letter of the last name of the author of the book Max chose?

17 Visualize Spatial Relationships

Many problem-solving strategies revolve around organizing information spatially. Spatial organization accentuates different aspects of a problem so that our strong visual sense and visual experience become more involved in solving the problem.

Graphs are one form of spatial organization. Instead of—or in addition to—organizing data in a chart, you can use a graph to organize data in a picture. Organized this way, the data are worth a thousand words. Higher-level thinking skills like graphing data and interpreting graphs are becoming more important as technology takes more of the burden of computation.

Besides making a graph, other strategies that rely heavily on some form of spatial organization include drawing a diagram, using physical representations, and drawing Venn diagrams. Scale drawings also appear in this chapter.

This chapter and strategy abound with practical implications. Each of the problems in the first part of the chapter could be realistically solved by graphing, with the possible exception of the Phone Calls problem (why not just call the phone company or look in the phone book to find out its rates?).

Problem Set A includes a number of problems that you might view as trigonometry problems. Encourage your students to make scale drawings instead of using trigonometry. The experience of using drawings will carry over into trigonometry problems. By making scale drawings, students will not only form mental images of the problem but will also develop organizational schemes for such problems. The skills and experience of drawing appropriate diagrams will give the students insights to complement a trigonometric approach.

Notes on Text Problems

CHICKEN NUGGETS

The most convincing solutions to this problem tend to be those where graphs and algebra are combined. We believe something more readily if we can see it, and the graph makes that possible.

PHONE CALLS

You may wish to discuss step functions with the students. The phone call charges function is an excellent, easy-to-understand example of a step function that comes from real life.

VACATION

The main feature of this problem is that graphs can be used to extend data, both between known data points and beyond the boundary of known data points.

Note on Problem Set A, Version 2

PIZZA

Even though a line graph does not represent real-world pizza sizes, a line graph on each of these may help students see the pattern.

Good Quiz Problems

We recommend no quiz on this chapter.

Text Problems

CHICKEN NUGGETS

A local fast-food vendor sells chicken nuggets for the following prices:

SERVING SIZE	PRICE
6 nuggets	$2.40
10 nuggets	$3.60
15 nuggets	$5.10
24 nuggets	$7.80

Draw a graph of this information. Then answer the following questions:

a. What is the equation for this graph?

b. What is the slope for the graph? What is the real-world significance of the slope in relation to the price of chicken nuggets?

c. What is the y-intercept for this graph? What is the real-world significance of the y-intercept?

d. How much would it cost to buy a serving size of 50 chicken nuggets?

e. Another restaurant sells 13 nuggets for $4.25 and 20 nuggets for $6.75. Use the graph to find out whether these are good deals compared to those of the restaurant above. Then use the equation for the graph to check your answer.

©2004 Key College Publishing, *Instructor Resources: Crossing the River with Dogs,* Johnson/Herr/Kysh

Text Problems *(continued)*

PHONE CALLS

During spring break, Josh went on vacation for five days. Aletta called him each of the five days he was gone. On two days she called during the day at the day rate. She made the other three calls at the evening rate.

TIME (IN MINUTES)	COST
11	$2.26
17	$4.57
6	$1.26
22	$4.46
7	$2.07

Determine the connect fee and the cost per minute for the two rate schedules Aletta called under: day rate and evening rate. (A connect fee is the charge levied the instant a phone conversation begins.) Note that both the day rate and the evening rate are linear relationships. Work this problem before continuing.

VACATION

The Family family wants to take another vacation. They have decided to drive their van to a destination 600 miles away. If they drive at a reasonable speed, how much time should they set aside for the trip? Plot a graph that shows various driving speeds (in miles per hour) on the *x*-axis and driving time (in hours) on the *y*-axis. Work this problem before continuing.

Text Problems (continued)

FAT CONTENT

Many ads for food indicate either the percentage of fat the food contains or to what percentage the food is fat-free. For example, an ad might say "Our burgers are 90% fat-free," or perhaps "Our lean hamburger is only 15% fat." A person reading these ads might assume that the percentage of calories from fat is also only 10% in the first case and 15% in the second case. But this assumption wouldn't be accurate, because the ads indicate the percentage of fat by *weight,* not in *calories.*

Fat has 9 calories per gram. Carbohydrates and protein each have only 4 calories per gram. In addition, virtually all foods have moisture content (as well as minerals, and so on.), which contains 0 calories per gram. For example, 2% milk (2% of the weight of the milk comes from fat) is about 89% water. When this water weight is taken into account, about 34% of the calories come from fat.

Draw a graph that shows a food's percentage of fat by weight on the *x*-axis and its percentage of fat in calories on the *y*-axis. Assume that this food is 50% water and that the remaining percentage of the food represents carbohydrates and protein. Then use your graph to find what percentage of fat by weight gives 50% of the calories from fat. Work this problem before continuing.

DUCKS AND COWS

Farmer Brown has ducks and cows. The animals have a total of 12 heads and 32 feet. How many ducks and how many cows does Farmer Brown have? Solve this problem with a graph before continuing.

MAXIMUM AREA

A farmer with 100 feet of fencing wants to build a rectangular garden. What should the dimensions of the garden be in order to enclose the maximum area? Work this problem before continuing.

Text Problems *(continued)*

DUCKS AND COWS

Farmer Brown has ducks and cows. The animals have a total of 12 heads and 32 feet. How many ducks and how many cows does Farmer Brown have? Read the explanation of how to solve this problem with a special type of scale drawing on page 450 of the text.

PATIO FURNITURE

A rectangular patio measures 18 feet by 12 feet. On the patio is a rectangular lounge chair that measures 2 feet by 4 feet 3 inches and a circular picnic table that measures 5 feet 8 inches in diameter. Make a scale drawing of the patio before continuing.

MAYDAY

"Mayday, Mayday!" The call startled Ned, who was stationed in the Coast Guard office. He immediately radioed: "Coast Guard here. What is your position? Over."

"I'm not sure. We left the port at Miami at 7:30. We sailed due southeast for 2 hours at 35 knots. Then we turned about 30° to starboard [right] and sailed for 4 hours at 25 knots. Then we lost our engines, and we have been adrift for about an hour and a half. We would have called earlier, but our radio was out. Can you send us some help?"

Ned replied, "I'll work out your position and send out a chopper right away. Over." Ned knew that the ocean current at that time of day was approximately 5 knots due south. A knot (kn) is 1 nautical mile per hour. One nautical mile (NM) is about 1.15 land miles. The helicopter speedometer measures land miles per hour. The likely speed of the helicopter is 80 miles per hour (that is, land miles). In what direction should Ned send the helicopter, and how many minutes will it take it to get to the stranded boat?

Make a scale drawing and solve this problem before continuing.

Problem Set A, Version 2

I. **DENISE'S AUTO REPAIR**

Sometimes auto repairs take a while because the mechanic has to test a car, let it cool, then investigate the problem. Other times repairs are delayed while the mechanic waits for parts to be delivered. When Denise opened her auto shop, she tried to keep a few cars in the shop so that she always had at least one to be working on. Some customers, she found, were very demanding about getting their cars back right away, while others, though they needed their cars, were very patient about waiting. Denise decided to do things a bit differently: She decided to charge those customers who wanted their cars right away more money and to charge those who were more patient less money. At the time, she wanted to make an average of about $20 per hour. The customers who could wait four days before getting their cars back would be charged only $15 per hour. Those who wanted their cars back within one day (24 hours) would be charged $60 per hour. At these rates, Denise figured that she could hire people to do things around her house. For example, if someone was really willing to pay $144 an hour in return for the guarantee of repair within 10 hours, Denise would in turn hire a couple of neighbors to watch her kids after school, cook dinner, clean house, and so on. She could pay them generous rates and still be making some pretty good money for herself. So Denise made up a table:

Denise's Auto Repair

Hourly Rates

Choose the schedule. I will guarantee your
car back within that amount of time.

Repair within	Hourly charge
6 hours	$240
10 hours	$144
18 hours	$ 80
24 hours	$ 60
30 hours	$ 48
36 hours	$ 40
96 hours	$ 15

---

a. Graph the information and estimate what she charged per hour for someone who wanted his or her car back within 12 hours.

b. Estimate how much she would charge for someone who wanted his or her car back within 60 hours.

c. Estimate how much she would charge someone who wanted his or her car back within 72 hours.

TEACHING RESOURCES

©2004 Key College Publishing, *Instructor Resources: Crossing the River with Dogs,* Johnson/Herr/Kysh

Problem Set A, Version 2 *(continued)*

2. **RUNNING ON M. T.**

Marie Taymor (M. T.) runs for health and also trains for middle distance races. Graph these times for M. T.'s various exercise distances:

Time (minutes)	Distance (miles)
5.5	1
11.5	2
18	3
27	4
35	5
64	8
108	12
150	15

a. What would be a reasonable time for M. T. to run 18 miles in?

b. If M. T. had timed herself running for 70 minutes, how far do you think she would have run?

3. **KATHY'S CATERING SERVICE**

Kathy is considering opening a catering service. In the beginning, she will only be catering to small groups. To get an idea what to charge, she called other caterers with a comparable menu to get their rates for various-sized groups.

Jake's Catering

Number of people	Price/person
40–49	$11.95
50–69	$10.95
70 and up	$10.59

Sybil's Slicery

Number of people	Price/person
40–59	$11.50
60–79	$10.50
80–99	$ 9.99
100 and up	$ 9.50

What should Kathy charge for groups of 10 to 19, for groups of 20 to 29, and for groups of 30 to 39?

Problem Set A, Version 2 *(continued)*

4. **GROUND TURKEY**

If uncooked ground turkey meat is kept at 35°, it is estimated that it will stay fresh for about 20 days. On the other hand, if it is kept at 90°, it is estimated to stay fresh for 6 hours—after that, the bacteria count will be dangerously high. Graph the following information:

TEMPERATURE	DAYS
40°	8
50°	4
70°	1

a. How long would you expect the meat to stay fresh if it were stored at 80°?

b. How long would you expect the meat to stay fresh if it were stored at 30°?

c. At what temperature would you expect it to stay fresh for 5 days?

5. **SELLING SODAS AT THE PARADE**

Darlene decided to sell sodas at the parade. She paid $1.20 for each six-pack and $1.50 for ice. She is charging 35¢ per soda. What is her break-even point?

6. **PIZZA**

The prices for a gourmet pizza at a local pizza parlor are listed below. Sizes are given in terms of the diameter, measured in inches.

	Diameter	Price
Personal	6 in.	$ 4.50
Small	10 in.	$ 8.00
Medium	14 in.	$11.00
Large	20 in.	$14.00

Graph this information on two graphs. On one graph, plot price versus diameter. On the other graph, plot price versus area. (*Note:* Plan ahead; part c will refer to a 24-inch pizza, so make sure your graphs can accommodate that size.)

(This problem continues on the next page.)

TEACHING RESOURCES

©2004 Key College Publishing, *Instructor Resources: Crossing the River with Dogs,* Johnson/Herr/Kysh

a. Does a pattern appear in the graph of price versus diameter? Describe it.

b. Does a pattern appear in the graph of price versus area? Describe it.

c. Suppose you were the manager and you decided to offer a family-size pizza that would be 24 inches in diameter. Using the price-versus-diameter graph, what price would you expect to charge? Using the price-versus-area graph, what price would you expect to charge? If you were the manager, what price would you truly charge and why?

7. THE CHICKEN COOP

The Family family just acquired 240′ of chicken wire with which to make an outside pen for their chickens. The pen will be rectangular in shape, and the wire must cover only three sides, because the existing coop will be the fourth side. Building in a rectangular shape, they want to make it the largest size possible (in area). One possibility for the size of the pen is 20′ by 200′. What should the dimensions be to make the area as large as possible?

8. LAUNCHING ROCKETS

Melissa and her friends Janet, Delbert, and Sharla are launching rockets. Janet's parachute got stuck in a tree. She is going to use a pole made out of plastic irrigation pipes to retrieve it. About 40 feet away, Melissa measured the angle of elevation for the rocket and parachute to be 33°. She is standing at the point directly beneath the parachute. How long a pole does she need to retrieve her rocket and parachute?

9. ROCKET TO THE MOON

Sharla shot off her rocket. Delbert and Melissa were in charge of figuring out the height. The line from where Delbert was standing to the rocket made a 73° angle from the ground. From where Melissa was standing, 50 feet behind Delbert, it was a 66° angle. How high did the rocket go?

10. DELBERT'S LAUNCH

When Delbert shot off his rocket, the friends tried different vantage points. Melissa stood about 100 yards away from Sharla with the launching pad between them on a straight line. The rocket went pretty much straight up. Sharla measured a 43° angle from the ground to the rocket, and Melissa measured a 35° angle. How far up did Delbert's rocket go?

©2004 Key College Publishing, *Instructor Resources: Crossing the River with Dogs,* Johnson/Herr/Kysh

Problem Set A, Version 2 *(continued)*

11. **TRAVERSING THE TRIBUTARY WITH CANINES**

Rover and his dog pals are standing on the other side of the river, starting to wade in to swim across. You are slightly upriver from the dogs, and your line of sight to the dogs makes an 80° angle with the bank. You then walk about 30 yards downstream and sight again. This time you are downriver (compared to the dogs) and it is about a 65° angle. How far across the river must the dogs swim?

12. **FLYING PAPER AIRPLANES**

Melissa and her friends decide to fly their paper airplanes out of the window of a tall building. Delbert is part of the ground crew, and Janet is the launch director. Delbert sights the window at a 75° angle from the ground, then backs up 25 feet and sights it again, this time at a 64° angle. How high is the window?

13. **JANET'S PAPER AIRPLANE**

See the previous problem, Flying Paper Airplanes. Janet, one of Delbert's friends, launches the first paper airplane. She watches where it lands. From her perspective, the angle from the side of the building to where it lands is about 68°. How far away from the building does the airplane land?

TEACHING RESOURCES

©2004 Key College Publishing, *Instructor Resources: Crossing the River with Dogs*, Johnson/Herr/Kysh

Problem Set B, Version 2

1. BOXCARS

Jared works switches in a train yard. He has to position some of the train's cars in the right order. He has an engine, a boxcar, a flatcar, a gondola car, and a tanker, in that order. He needs to change the order to engine, tanker, boxcar, gondola car, and flatcar. He has only this section of track to work with:

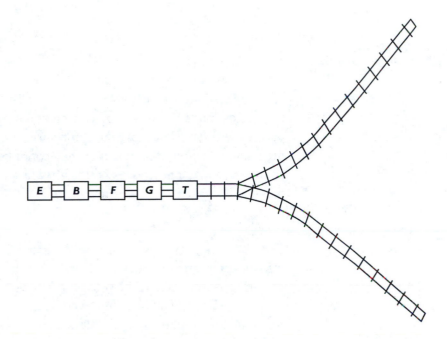

Note that Jared cannot convince the cars to get off the tracks and move into position. Rather, he must back up, unhitch or hitch, and drive forward to accomplish his task. How can he do it most efficiently?

2. THEY'RE TWINS

Sisters Lynn and Laurie play on the QueenAirs basketball team. Last month they each scored the same number of points—amazing! Together, they had the same number of one-point free throws as two-point field goals, and this number was also the number of three-point field goals. Lynn had three times as many free throws as she had two-point field goals. Laurie had five times as many two-point field goals as she had three-point field goals. They scored less than 200 points total. How many of each type of score did they each make?

©2004 Key College Publishing, *Instructor Resources: Crossing the River with Dogs*, Johnson/Herr/Kysh

Problem Set B, Version 2 *(continued)*

3. RUDY'S ROOT BEER

Rudy likes his root beer. Sometimes he has it in the morning, sometimes he likes it in the afternoon. When he drinks it in the morning, he puts ice in it. When he drinks it in the afternoon, he puts ice cream in it. On vacation, Rudy had root beer on 19 different days. If there were 8 afternoons with no root beer and 15 mornings with no root beer, what is the fewest number of days his vacation could have been?

4. DIGGING A TUNNEL

Two crews digging a tunnel through a mountain start on opposite sides of the mountain, 120 feet apart. The first crew is making about 5 feet of progress per day, and the second crew is making about 4 feet of progress per day. After four days, the surveyor discovers that they are digging parallel tunnels, which, if they continue, will be about 35 feet apart. At this point each crew turns toward the other and continues digging. When will the two crews meet?

5. TOASTER SERIAL NUMBER

Crustee Toasters issues serial numbers made up of two digits, then two letters, then two more digits and a letter. For example, 23UF51T is a valid serial number. The last letter advances consecutively with each new toaster, so 23UF51T is followed by 23UF51U, then 23UF51V, and so on, all the way to 23UF51Z. Once Z is reached, the last letter resets to A and the number in the next column to the left increases by 1. So the toaster after 23UF51Z would carry the number 23UF52A. After toaster 98BA76Z, how many more toasters will be made before all digits and letters in the serial number are different again?

TEACHING RESOURCES

©2004 Key College Publishing, *Instructor Resources: Crossing the River with Dogs*, Johnson/Herr/Kysh

Problem Set B, Version 3

1. BEDTIME

It was late and I was tired and the kids were still up hopping around. I told them to get into bed, but my daughter Alyse said, "It's not time yet." I replied that it was 9:58, but she said, "No, it's 8:56." (She was looking at the clock upside down.) How many times in a day can a regular 12-hour digital clock be read upside down and still give a valid time? (Ignore the space where the colon is or should be.)

2. CLASSIC CAR CLUB

A note in the newsletter of a classic car club described the 16 cars owned by members of the club. Every car is described with at least one of the following statements. Among the 16 cars, there were

1. At least one red convertible.

2. More blue sports cars than red sports cars.

3. Fewer convertibles than sports cars.

4. More blue convertibles than blue sports cars.

5. No convertible sports cars.

Unfortunately, one of cars was destroyed in a garage fire. However, the note in the newsletter was still accurate. What is the description of the car that was destroyed?

3. BACK IN SHAPE

Jody seriously needed to get back in shape. She also had a lot of yard work to do, so she decided to combine the two needs into one project. She decided to dig and move wheelbarrows full of dirt. On the first day, she dug and moved $1/2$ a wheelbarrow full. On the second day, she dug and moved $1/3$ of a wheelbarrow full and then $2/3$ of a wheelbarrow full. That was enough work. On the third day, she dug and moved dirt by fourths of a wheelbarrow: She moved $1/4$ as a warm-up, and then $2/4$ and $3/4$ as her exercise. She continued doing this, building up the denominator and starting with light loads as warm-ups, then coming close to moving one full wheelbarrow on the last move, but never quite making it. The fourth day she dug and moved $1/5$, $2/5$, $3/5$, and $4/5$ wheelbarrow loads. She did this for a total of 47 days. How many wheelbarrow loads of dirt did she dig and move in that time?

Problem Set B, Version 3 (continued)

4. **THE WEDNESDAY CLUB**

The Wednesday Club was named for the day of the week it met. Soon after it formed, though, the Wednesday Club started another meeting on Tuesday for members and prospective members who could not make the Wednesday meetings. Since the club was so flexible, it soon expanded to Thursday, and some members who came to the Wednesday or Tuesday meetings started coming to the Thursday meetings. Others went to the Wednesday, or Tuesday, or Thursday meetings only; some went to two meetings; and some went to all three. Well, obviously, a club this good couldn't be kept a secret, and membership and demand expanded the club to all days of the week.

On some days, different sections of the club met. For example, the Wednesday Club Book Section met Friday at 1:00, and the Wednesday Club Skateboarders Section met Tuesday at 2:00. The Wednesday Club Movie Section met on Mondays at 4:00, and so on—you get the picture. (Occasionally the Wednesday Club got together on Friday to watch a videotape of *Monday Night Football: Thursday Edition*.)

As it turned out, in one particular week every member attended at least one meeting. At least one member attended each meeting, and no identical set of members attended any two of the same meetings during the week. There were 8191 meetings during that week. What is the minimum number of members needed to accomplish this amazing feat?

5. **WOLF MOUNTAIN SKIER'S PLAN**

Weekend skiers at Wolf Mountain can register under a frequent-skier program to receive 3 points toward a free lift ticket. Weekday skiers receive 10 points toward a free lift ticket. Julie, Justin, and Jamie together received 101 points this last season. Each skied the same number of days, but they didn't all go on the same days. How many weekdays and how many weekend days did they ski, collectively?

TEACHING RESOURCES

©2004 Key College Publishing, *Instructor Resources: Crossing the River with Dogs*, Johnson/Herr/Kysh

Problem Set B, Version 4

1. HUESOS DE GALLO

My five-letter word has exactly two letters in common with each of the following words. What is my word?

BLAZE   GRAVY   QUILT   GHOST   CHUMP

2. WOLFIES ON THE RUN

The offense for the Wolfies is bad. They scored a touchdown in each of 13 different football games, but they never scored a touchdown in both halves of any game. Touchdowns in the first half were followed by scoreless second halves, and touchdowns in the second half occurred only if the first half was scoreless. There were 11 scoreless first halves and 12 scoreless second halves in all. How many games did the Wolfies play this season?

3. WHEAT FIELD

County Road 12 runs straight north-south from Brighton to Chester. Manjeet is on County Road 17 (which runs east-west), about 3 miles from the intersection of the two roads, and she's about to walk home. Her house is 5 miles south of the intersection. Manjeet cuts across the wheat field to a point on County Road 12 so that half of her trip is through the wheat field and half is on County Road 12. How far does she go using this route?

4. DIGITAL CLOCK

There is a mirror in my bathroom. If I look in the mirror while I am shaving, I can see my digital clock. It is a 12-hour clock. Sometimes the time that I see in the mirror causes me to panic because I think it is the actual time, when in fact it is a different time. For example, 852 when reflected in the mirror becomes 528, which is a legitimate time, but not the real time. (I can't see the colon that separates the hour from the minutes.) How many times in a 24-hour day does the clock in my mirror read a time that could be a real time but is not the actual time?

5. **PERFECT SHUFFLE**

A perfect shuffle is defined as follows: You separate the deck of cards into two equal piles with the top half of the deck in the left pile. Then you intersperse one card from each pile throughout the whole deck. So, after the shuffle is over, the top card of the left pile is on top, with the top card of the right pile under it. The third card is the second card from the left pile, the fourth is the second card from the right pile, then the third card from the left pile, the third card from the right pile, and so on, throughout the deck. How many perfect shuffles does it take to get a 52-card deck back into the order it was in before you started shuffling?

Sample Midterms and Final Exam Options

Sample Midterms

The next pages contain sample midterms. We feel that giving a midterm is a good idea, but you should allow the students enough time, and you shouldn't expect them to solve all of the problems completely. As you will see in these samples, many of the problems do not ask for complete solutions. Rather, they ask for specific parts of the solution. In this way you can see if your students understand each strategy. The midterms that follow took 40 to 60 minutes for the students to do. We allowed students to use calculators on the midterm. This midterm was given at the beginning of week 11, after Chapters 1–9 had been completed.

Sample Midterm I

Name _____

Directions: Make sure you follow the instructions for each problem. If you don't, you will run out of time. Each part is worth 10 points. Do your work on a separate sheet of paper.

Part I: Easier Related Problems

1. The book lists six ways to make a problem easier. List three of these.

2. For the following problem, **state and solve** three easier related problems. You do **not** have to solve the actual problem.

   ### DIFFERENCES
   Find the difference between the sum of the first 600 multiples of 3 and the sum of the first 600 odd numbers.

Part II: Subproblems

3. For the following problem, state all the subproblems that you would have to solve in order to solve the problem. When a problem says, "State a subproblem," write a complete sentence. **You do *not* have to solve any of your subproblems, nor do you have to solve the problem.** You don't even have to explain how to solve any of the subproblems.

   ### JACKET SALE
   Leigh went to the mall to buy a new jacket. The one he liked was marked at $84, but it was sitting on a table with a sign that read, "Take $12 off the price of everything on this table." When he bought the jacket, the clerk gave him the $12 discount. Then he decided to use the 15% off coupon in his wallet. Then the sales tax of 7.25% was added on. How much change did Leigh receive from his hundred-dollar bill?

Part III: Guess-and-Check

4. Set up a guess-and-check chart with good, clear labels for the following problem. Make **three** guesses and then *stop*. Your guesses must bracket the answer, so you should have at least one low guess and one high guess. If you have to make more than three guesses to achieve this, that is okay. It is not necessary to solve the problem. In fact, don't solve it because you will run out of time for other problems.

   ### A VERY LONG JOURNEY
   Mallory drove at a speed of 58 miles per hour from her house to her sister Morgan's house. The two of them then drove at a speed of 63 miles per hour to Santa Barbara to visit their friend Tracy. Mallory spent a total of 7 hours in the car (from her house to Morgan's house to Tracy's house). The total distance Mallory traveled (from her house to Morgan's house to Tracy's house) was 430 miles. How far is it from Mallory's house to Morgan's house?

Part IV: Matrix Logic

5. You are going to solve a matrix logic problem involving the following four categories:

first names, last names, favorite baseball team, and favorite TV show

Set up the matrix logic chart you would use to solve the problem. Use good, clear labels. You don't have to make up names; **we are just looking for the categories** in the chart. Obviously you can't actually solve the problem because you have no clues or any of the actual names, and so on.)

Cross Correlating—Problems 6 and 7
In each matrix below: **What** can you conclude? Make **one** additional mark in the matrix. **Explain why.**

Note: There would be a lot of other lines in the chart for other people, and so on, but we didn't show those. We're **not** talking about placing an **X** in the cells for the other first and last names since Megan is Bram.

Also note: **O** means yes and **X** means no.

6.

		Last names				Cars	
		Bram				2-door	
First names	Megan	O					
Cars	2-door	X					

7.

		Last names				Pets	
		Velk				dog	
First names	Blane	O				O	
Pets	dog						

Part V: Unit Analysis

8. **Solve** this problem by unit analysis. Show all the units and fractions in the setup to this problem. *Note:* 1 mile = 5280 feet   3 feet = 1 yard

RUNNING TO CLASS
TJ runs at a speed of 7 miles per hour. How many seconds will it take him to get from the cafeteria to Weaver Hall, a distance of 62 yards? (Round your answer to the nearest one-hundredth of a second.)

Sample Midterm 2

Name _____

Directions: Make sure you follow the instructions for each problem. If you don't, you will run out of time. Each part is worth 10 points. Do your work on a separate sheet of paper.

Part I: Easier Related Problems

1. The book lists six ways to make a problem easier. List three of these ways.
2. For the following problem, **state and solve** three easier related problems. You do **not** have to solve the actual problem.

MULTIPLES OF 4
What is the sum of the first 2000 multiples of 4?

Part II: Subproblems

3. For the following problem, state all the subproblems that you would have to solve in order to solve the problem. When a problem says, "State a subproblem," write a complete sentence. **You do *not* have to solve any of your subproblems, nor do you have to solve the problem.** You don't even have to explain how to solve any of the subproblems.

USED CAR
Three years ago, you bought a new car for a list price of $20,000. You also paid sales tax of 7.5%. The car is now worth 25% less than its original list price. Your brother-in-law doesn't know anything about cars, so you managed to sell the car to him for $400 more than the car is currently worth. What is the difference between the price at which you bought the car and the price at which you sold it?

Part III: Guess-and-Check

4. Set up a guess-and-check chart with good, clear labels for the following problem. Make **three** guesses and then ***stop***. Your guesses must bracket the answer, so you should have at least one low guess and one high guess. If you have to make more than three guesses to achieve this, that is okay. It is not necessary to solve the problem.

TRIKE COLLISION
Kristin and Cat are little girls that each own a cool trike. They decide to play chicken in the street. They start at the opposite ends of a long street 400 feet apart and ride toward each other. Kristin rides at 5 ft/sec. Cat rides at 8 ft/sec. In how many seconds will they meet (and crash into each other)?

©2004 Key College Publishing, *Instructor Resources: Crossing the River with Dogs,* Johnson/Herr/Kysh

Part IV: Matrix Logic

5. You are going to solve a matrix logic problem involving the following categories:

   first names, last names, favorite movies, and favorite sports

   Set up the matrix logic chart you would use to solve the problem. Obviously you can't actually solve the problem because you have no clues or any of the actual names, and so on. Use good, clear labels.

Cross-Correlating—Problems 6 and 7

In each matrix below: **What** can you conclude, and **why**? Make an additional mark in the matrix. Explain why.

Note: There would be a lot of other lines in the chart for other people, and so on, but we didn't show those. We're **not** talking about placing an X in the cells for the other first and last names since Lance is Kensey.

Also note: O means yes and X means no.

6.

	Last names	Sports
	Kensey	golf
First names Lance	O	
Sports golf	O	

7.

	Last names	Pets
	Riev	bird
First names Andra	O	X
Pets bird		

Part V: Unit Analysis

8. **Solve** this problem by unit analysis. Show all the units in the setup to this problem.
   Note: 1 mile = 5280 feet   3 feet = 1 yard

   ONE-HUNDRED-YARD DASH

   If Erica runs at a constant speed of 16 miles per hour, how many seconds does it take her to run 100 yards? (Round your answer to the nearest one-hundredth of a second.)

©2004 Key College Publishing, *Instructor Resources: Crossing the River with Dogs*, Johnson/Herr/Kysh

Answers to Sample Midterms

1. List any three of the six ways mentioned in the book. The exact wording is not important.
   a. Use a number instead of a variable.
   b. Use a smaller or easier number to develop the process for solving the problem.
   c. Do a set of easier examples and look for a pattern.
   d. Do an easier example and figure out how to use the same process to solve the harder problem.
   e. Change, fix, or get rid of some conditions.
   f. Eliminate unnecessary information.

2. Any three easier related problems, stated and solved. Examples follow.
   Find the difference between the sum of the first four multiples of 3 and the sum of the first four odd numbers.
   $3 + 6 + 9 + 12 = 30$
   $1 + 3 + 5 + 7 = 16$
   $30 = 16 - 14$
   Find the difference between the sum of the first five multiples of 3 and the sum of the first five odd numbers.
   $3 + 6 + 9 + 12 + 15 = 45$
   $1 + 3 + 5 + 7 + 9 = 25$
   $45 - 25 = 20$
   Find the difference between the sum of the first two multiples of 3 and the sum of the first two odd numbers.
   $3 + 6 = 9$
   $1 + 3 = 4$
   $9 - 4 = 5$

3. List the subproblems. The order may be different from what follows.
   How much was the jacket after the $12 discount?
   How much is 15% of the discounted price?
   What is the new discounted price?
   How much is the sales tax?
   What is the final price including sales tax?
   How much change will Leigh receive?

4. A good guess-and-check chart should be set up, with good, clear labels and a rating column. All calculations must be correct. There must be one high guess and one low guess.

5. Any six-chart matrix is fine, set up similar to the one on page 91 of the text.

6. Mark an **X** in the Megan-2-door space. Megan is Bram, and Bram does not drive a 2-door; therefore, Megan does not drive a 2-door.

7. Mark an **O** in the Velk-dog space. Blane's last name is Velk, and Blane owns a dog; therefore, Velk owns a dog.

8. 18.12 seconds

1. List any three of the six ways mentioned in the book. The exact wording is not important.
   a. Use a number instead of a variable.
   b. Use a smaller or easier number to develop the process for solving the problem.
   c. Do a set of easier examples and look for a pattern.
   d. Do an easier example and figure out how to use the same process to solve the harder problem.
   e. Change, fix, or get rid of some conditions.
   f. Eliminate unnecessary information.

2. Any three easier related problems, stated and solved. Examples follow.

   What is the sum of the first three multiples of 4?

   $4 + 8 + 12 = 24$

   What is the sum of the first five multiples of 4?

   $4 + 8 + 12 + 16 + 20 = 60$

   What is the sum of the first six multiples of 4?

   $4 + 8 + 12 + 16 + 20 + 24 = 84$

3. List the subproblems. The order may be different from what follows.

   How much was the sales tax on the car?

   What was the original price of the car, including the sales tax?

   How much is 25% of the original list price?

   What is the car worth today?

   How much will your brother-in-law pay for the car?

   What is the difference between the amount you received for the car and the amount you paid?

4. A good guess-and-check chart should be set up, with good, clear labels and a rating column. All calculations must be correct. There must be one high guess and one low guess.

5. Any six-chart matrix is fine, set up similar to the one on page 91 of the text.

6. Mark an **O** in the Lance-golf space. Lance's last name is Kensey, and Kensey plays golf; therefore, Lance plays golf.

7. Mark an **X** in the Riev-bird space. Andra is Riev, and Andra does not own a bird; therefore, Riev does not own a bird.

8. 12.78 seconds

Final Exam Options

Y̶ou may or may not choose to give a final exam, depending on whether you think it would be useful and whether your finals schedule allows you to give students adequate time. You and your students should not view this final in the same way you might view final exams in traditional courses. Whereas many final exams are used to evaluate students' accumulation of knowledge over a semester and account for a large part of students' grades, this problem-solving final is designed to provide students with a final chance to apply what they've learned to a new situation. Most problems on the final are different (some are quite difficult) from those students have solved before, and students will be solving problems under a time constraint.

Because students have worked in groups throughout the course, we allow them to work in groups on the final. (There's no way to allow students enough time to do a final unless they work in groups. If you don't want to give a group final, we suggest you don't give a final.) In this way too, the final is different in that students receive a group grade.

We count the final as, at most, 10% of a student's grade so that students see evaluation as an important, but not overriding, goal of the final; they take it seriously, but they don't feel the same level of pressure they might associate with finals in other classes. In fact, some students look forward to the final as a chance to "show their stuff" and to see for themselves what they've gained from the course. Many students have told us that the final was a fun way to end the semester.

Give students several days before the final to form groups of four or five people with whom they want to work on the final. As groups work on the final, they'll find their own efficient ways of working together, but you might suggest students start by tackling the problems in pairs and presenting a solution to the rest of the group to check the work. It should be obvious to students that every group member has a stake in questioning and checking the work of their peers.

This book does not include ready-to-copy final exams that you can give to your students. Instead, it provides families of problems from which you can choose. We recommend using six problems, of which students have to do five in a two-hour exam period. You'll probably want to add your own directions for the final. The directions we use for the final when we give it are presented on the next page.

©2004 Key College Publishing, *Instructor Resources: Crossing the River with Dogs*, Johnson/Herr/Kysh

You may use this paper to do scratch work, but any work that you want me to read for credit must be on the group paper, along with all of the names of the members of your group. Write up one solution to each problem for your group. The problems are worth 10 points each and will be graded the way problem sets are graded, so be sure to show and explain all work and strategies used.

You have to do only five of the six problems. Do a good job on those five. If you have time, you may write up the other problem for 5 points of extra credit. If you do turn in all six problems, clearly indicate which problem should be considered extra credit. Relax, have fun, and work together.

When you prepare your final exam, create a version with your directions and all of the problems on one or two pages, then make enough copies for individual students. Also create a version with one problem to a page and space at the top of each page for four or five group-member names. Each group gets one copy of this six-page version. They will turn in their work on these pages. If you copy each problem on different-colored paper, you'll find them easier to sort and you can grade every group's solution to one problem at a time.

The problems are arranged into families, which are separated by lines across the pages. Don't pick more than one problem from a family. The idea is that you can pick a different problem from a given family the next time you give a final. The problems are also rated by difficulty on a scale of 1 to 4, with 1 being easy and 4 being hard. Try to pick a mixture of difficulty levels. Don't tell students what the difficulty levels for the problems are.

Strategies are suggested for each problem, but don't suggest strategies to the students. The strategies are here for you so that you can pick a variety of strategies for the final-exam problems.

RIVER CROSSING 1

A group of 24 adults and 36 children come to a river they would like to cross. They find a small boat that will hold 1 adult or 2 children. Everyone is able to row the boat. How many trips will it take for everyone to get across the river?

Strategies: manipulatives, acting it out, patterns, easier related problems, maybe finite differences (although finite differences aren't necessary)

Difficulty: 2

Answer: 165 trips

RIVER CROSSING 2

A group of 34 adults and 27 children come to a river they would like to cross. They find a small boat that will hold 1 adult or 2 children. Everyone is able to row the boat. How many trips will it take for everyone to get across the river?

Strategies: manipulatives, acting it out, patterns, easier related problems, maybe finite differences (although finite differences aren't necessary)

Difficulty: 2

Answer: 187 trips

Note: $\text{Trips} = 4a + 2c - 3$

LAKE CROSSING 1

A group of 37 adults and 55 children want to cross a lake. They find a small boat that will hold 2 adults, or 3 children, or 1 adult and 1 child. Every person is capable of operating the boat alone. What is the smallest number of trips it will take to get everyone across the lake? (Count trips as one-way trips.)

Strategies: manipulatives, acting it out, patterns, easier related problems, maybe finite differences (although finite differences aren't necessary)

Difficulty: 3

Answer: 107 trips

©2004 Key College Publishing, *Instructor Resources: Crossing the River with Dogs,* Johnson/Herr/Kysh

A group of 41 men and 63 women want to cross a lake. They find a small boat that will hold 2 men, or 3 women, or 1 man and 1 woman. Every person is capable of operating the boat alone. What is the smallest number of trips it will take to get everyone across the lake? (Count trips as one-way trips.)

Strategies: manipulatives, acting it out, patterns, easier related problems, maybe finite differences (although finite differences aren't necessary)

Difficulty: 3

Answer: 123 trips

CALCULATORS

There were 75 students in a precalculus class. Every student in the class had a calculator. Every calculator was either a Tasio or an Arkansas Instruments (AI). There were nine more AIs than Tasios. Four times as many students had graphing calculators as had nongraphing calculators. Ten students had nongraphing AIs. How many students had graphing Tasios?

Strategies: Venn diagrams, systematic lists, algebra, organizing information

Difficulty: 2

Answer: 28

PARTY

There are 48 people at a party. There are seven times as many adults as children. There are twice as many females as males. If there are 5 girls (female children) at the party, how many men (male adults) are there?

Strategies: Venn diagrams, systematic lists, algebra, organizing information

Difficulty: 1

Answer: 15

TOY CAR RACE

Dale, AJ, and Mario each have a battery-operated toy car. The cars always travel in a straight line at a constant rate of speed. They decided to have a race. Each car started at the same time at the beginning of a straight race course. When AJ's car crossed the finish line, it was ahead of Dale's car by 24 inches and was ahead of Mario's car by 32 inches. When Dale's car crossed the finish line, it was ahead of Mario's car by 10 inches. How many inches long is the race course?

Strategies: diagrams, working backwards

Difficulty: 2

Answer: 120 inches

LETTUCE-EATING SNAILS

Kate has three African snails named Sluggo, Pokey, and Lag. Each of the snails loves to eat lettuce. One day, Kate gave each snail the same number of heads of lettuce. The snails started munching away. Each snail ate at a constant rate of speed. When Sluggo finished all of his heads of lettuce, Pokey had 17 heads left and Lag had 26 heads left. When Pokey finished all of his lettuce, Lag had 12 heads left. How many heads of lettuce did each snail start with?

Strategies: diagrams, working backwards

Difficulty: 2

Answer: 68 heads

COUNTING VOTES

Al, George, Kathleen, and Chad counted ballots in Florida. They were each given a box of ballots. Each box contained the same number of ballots. Each person counted ballots at a constant rate. However, the rate at which each person counted ballots was different. When Al finished counting all of his ballots, George had 54 ballots left to count, Kathleen had 90 ballots left to count, and Chad had 106 ballots left to count. When George finished counting all of his ballots, Kathleen had 45 ballots left to count and Chad had 65 ballots left to count. When Kathleen finished counting all of her ballots, how many ballots did Chad have left to count? How many ballots were in each box to start with?

Strategies: working backwards

Difficulty: 2

Answer: Chad will have 24 ballots left. There were 270 ballots in each box to start with.

©2004 Key College Publishing, *Instructor Resources: Crossing the River with Dogs*, Johnson/Herr/Kysh

LOTSA NINES

What number times the number 434782608695652173913 gives all 9's for an answer?

Strategies: eliminating possibilities, patterns, working backwards, guess-and-check

Difficulty: 2

Answer: 23

A WHOLE BUNCH OF THREES

What number times the number 17543859649122807 gives all 3's for an answer?

Strategies: eliminating possibilities, patterns, working backwards, guess-and-check

Difficulty: 2

Answer: 19

BUNCHES OF NINES

What number times the number 3448275862068896551724137931 gives all 9's for an answer?

Strategies: eliminating possibilities, patterns, working backwards, guess-and-check

Difficulty: 2

Answer: 29

Start with the ten-digit number 4261518507. This number is considered the original number.

Divide 4261518507 by 2 and write down the remainder.
Divide 4261518507 by 3 and write down the remainder.
Divide 4261518507 by 4 and write down the remainder.
Divide 4261518507 by 5 and write down the remainder.
Divide 4261518507 by 6 and write down the remainder.
Divide 4261518507 by 7 and write down the remainder.

You should now have a set of six remainders, from dividing by 2, 3, 4, 5, 6, and 7.

Now, reverse a pair of adjacent digits in the original ten-digit number, say, the 7 and the 0. Your new number is 4261518570. Perform all the same divisions you did for the original number, and write down all the remainders. Note that some of the remainders are the same as they were for the original number and some are different.

Go back to the original number and reverse a different pair of adjacent digits. Again, perform the same divisions and list the remainders. Repeat this process for every possible pair of adjacent digits in the original number.

Surprisingly, for one of the pairs of adjacent digits, all the remainders are the same as those you found when you divided the original number. Switching which pair of adjacent digits caused this to happen?

Strategies: guess-and-check, easier related problems, patterns, eliminating possibilities

Difficulty: 3

Answer: 1 and 8

©2004 Key College Publishing, *Instructor Resources: Crossing the River with Dogs*, Johnson/Herr/Kysh

An accountant was sitting at her desk on a break, so she started playing around with her computer. She entered the number 5739216507 and divided it by 2, then noted the remainder. She also divided 5739216507 by 3 and noted the remainder. She also divided it by 4, then by 5 and then by 6, and noted the remainder each time. Finally, she divided it by 7 and noted the remainder. Then, since she was still on her break, she reversed a pair of adjacent digits in the original number and went through the whole process again, dividing by 2, 3, 4, 5, 6, and 7, noting the remainder each time. Then she again started with the original number and reversed another pair of adjacent digits and divided and noted the remainders. She did this for every possible pair of adjacent digits. Surprisingly, for one pair of adjacent digits all the remainders were the same as those reached with the original number. Which pair of adjacent digits caused this to happen?

Strategies: guess-and-check, easier related problems, patterns, eliminating possibilities

Difficulty: 3

Answer: The 9 and the 2

■ **MATH MAGIC**

A man has 11 cards. On each card, he has written a positive whole number. He then asks you to say any whole number from 1 to 2000. You tell him a number. He then shows you a certain number of the cards he is holding. The numbers on the cards he shows you will add up to exactly whatever number you said. This always works, for any number you say from 1 to 2000. What numbers are on the cards?

Strategies: patterns, systematic lists, eliminating possibilities, working backwards

Difficulty: 3

Answer: Cards should be numbered 1, 2, 4, 8, 16, 32, and so on, up to 1024. The last card can actually be less than 1024, as long as the sum of all cards is greater than or equal to 2000.

When the attendance sheet reached Mr. Dylan, he called out, in his usual way, "Has anyone *not* signed the attendance sheet?" As he asked this question, he thought to himself, "This must be the 265th time I've asked this. Actually, maybe it's the 265,265th time." Seven students came up to his desk to sign the attendance sheet, because they hadn't signed it when it was passed around the classroom.

As he waited for them to sign, Mr. Dylan thought, "Gee, I've made this announcement 265,265 times. Seven students didn't sign the sheet today. This is an eleven o'clock class, and today is Friday the thirteenth. Those numbers make me want to grab a calculator. Let's see, 265,265 divided by 7, then by 11, and then by 13 is equal to 265. That's amazing! Does this always work? Let me try it with another three-digit number, say 481. Punch in the six-digit number 481,481. Divide by 7, then by 11, then by 13 and the result is 481. Whoa! Why does this work?"

1. Explain why this calculation always works.

2. Write down a two-digit number three times to create a six-digit number, such as 393,939. What four numbers must you divide by consecutively to end up with your original two-digit number (in this case, 39)?

Strategies: guess-and-check, subproblems, eliminating possibilities, patterns, systematic lists, working backwards

Difficulty: 4

Answers:

1. $7 \times 11 \times 13 = 1001$, and when you multiply a three-digit number *abc* by 1001, you get a six-digit number that is of the form *abc,abc*.

2. When dividing *aba,bab* by *ab,* the answer is 10,101. This number factors into $3 \times 7 \times 13 \times 37$. So, divide the six-digit number *aba,bab* successively by 3, 7, 13, and 37 to get the two-digit number *ab*.

The matrix below is a 100-by-100 number matrix. Find the sum of all the numbers in the matrix.

6	10	14	18	22	26	...
15	25	35	45	55	65	...
24	40	56	72	88	104	...
33	55	77	99	121	143	...
⋮	⋮	⋮	⋮	⋮	⋮	

Strategies: patterns, finite differences, easier related problems

Difficulty: 3

Answer: 153,510,000

The matrix below is an 80-by-80 number matrix. Find the sum of all the numbers in the matrix.

36	44	52	60	68	76	...
41	49	57	65	73	81	...
46	54	62	70	78	86	...
51	59	67	75	83	91	...
⋮	⋮	⋮	⋮	⋮	⋮	

Strategies: patterns, finite differences, easier related problems

Difficulty: 3

Answer: 3,516,800

WHAT DO I NEED FOR A B?

In her math class, Mary Taylor has figured out that her average is 76%. Up to now, 900 points were possible in the class. From now until the end of the semester, 300 points will be possible, making a total of 1200 points possible for the entire semester. Mary really wants a B in the class, so she wants to figure out what her average has to be from now on in order for her to end the semester with an 80% average. What does her average have to be for the 300 points for her to get her B?

Strategy: subproblems

Difficulty: 1

Answer: 92%

BATTING AVERAGE 1

Suppose that during one baseball season, Jose Canseco had been at bat 472 times as of September 1 and had a batting average of .286. And suppose that his biggest fan figured Canseco would have 108 times at bat during the rest of the season. She hoped that Canseco would finish the season with a .300 batting average. What is the lowest batting average Canseco could maintain for the rest of the season to end up with a season average of at least .300?

Strategy: subproblems

Difficulty: 2

Answer: .361

BATTING AVERAGE 2

A baseball player has been at bat 439 times so far this season and has a batting average of .278. He figures that he will have 171 times at bat during the rest of the season. What batting average does he have to maintain for the rest of the season to end up with a season average of at least .300?

Strategy: subproblems

Difficulty: 2

Answer: .357

©2004 Key College Publishing, *Instructor Resources: Crossing the River with Dogs,* Johnson/Herr/Kysh

FREE-THROW PERCENTAGE

Suppose that Magic Johnson, former player for the Los Angeles Lakers, had shot 200 free throws after 37 games. At that time, he had a free-throw percentage of .835. There were 45 games remaining to be played in the season, and Magic figured he would shoot about the same number of free throws per game in the coming 45 games as he had been shooting in the previous 37 games. What free-throw percentage did he have to maintain for the 45 remaining games to end up with a season percentage of .900 (and fulfill the incentive clause in his contract)?

Strategy: subproblems

Difficulty: 3

Answer: .955

LAS ESTAMPILLAS DE SAN FERMIN

Princess Irina of the mythical country of San Fermin is considering printing only three denominations of stamps: 19 cents, 8 cents, and 5 cents. She wants you to advise her: What is the largest amount of postage that cannot be made using these denominations?

Strategies: systematic lists, patterns

Difficulty: 3

Answer: 22 cents

THE STAMPS OF THE TARNOVIK REPUBLIC

The Tarnovik Republic prints stamps in the denominations 17 cents, 9 cents, and 5 cents. What is the largest amount of postage that cannot be made using these denominations?

Strategies: systematic lists, patterns

Difficulty: 3

Answer: 21 cents

In the team sport called Shuttle, there are three ways to score points. A troi is worth 13 points, a gorn is worth 10 points, and a crusher is worth 6 points. What number represents the largest score that cannot be made in this game?

Strategies: systematic lists, patterns

Difficulty: 4

Answer: 27

PET STORE

I visited the pet store last week and found two kinds of pets for sale: rabbits for $5 and parakeets for $9. If I had $14 to spend, I could buy one rabbit and one parakeet. If I had $24 to spend, I could buy three rabbits ($15) and one parakeet ($9), which would amount to exactly $24. However, if I had $13 to spend, no combination of rabbits and parakeets would cost $13. Assuming that money is no concern, what is the largest amount of money that I could have and not be able to spend all of it on some combination of rabbits and parakeets?

Strategies: systematic lists, patterns

Difficulty: 4

Answer: $31

CCC

Suppose that three Civilian Conservation Corps members, Richard, Nancy, and John, had to build a rock wall. They had to move 21 boulders across a field. Seven of the boulders were large, weighing 25 pounds each. Seven of the boulders were medium-sized, weighing 20 pounds each. Seven of the boulders were small, weighing 15 pounds each. Each worker had a wheelbarrow. They agreed to share the task equally. Each person carried the same number of boulders and the same number of pounds. How did they manage to do this?

Strategies: diagrams, manipulatives, subproblems

Difficulty: 1

Answer: There are many possible solutions. Here is one solution: One person carries 3 large, 1 medium, and 3 small, while two people each carry 2 large, 3 medium, and 2 small.

©2004 Key College Publishing, *Instructor Resources: Crossing the River with Dogs*, Johnson/Herr/Kysh

MILK BUCKETS

You have two buckets: a 5-quart bucket and a 3-quart bucket. You need exactly 4 quarts of milk. You go to the store to get milk. It has a large vat of milk, from which you can fill up your buckets; you can also pour milk back into the vat. By pouring milk back and forth between the vat and your buckets, how can you measure out exactly 4 quarts of milk to take home? You may not use any other measuring device.

Strategies: diagrams, patterns, systematic lists

Difficulty: 1

Answer: There are many ways to do this.

MILK CONSUMPTION

Mrs. Robinson has three kids: Katherine, Benjamin, and Elaine. Katherine can drink a quart of milk in two days. Benjamin can drink a quart of milk in three days. Elaine can also drink a quart of milk in three days. The kids are expecting five guests for two days next week. How much milk will Mrs. Robinson need for the eight kids? (Assume that the milk preferences of three children are representative of the five guests' milk preferences.)

Strategies: systematic lists, diagrams, subproblems, unit analysis

Difficulty: 2

Answer: $6\frac{2}{9}$ quarts

MEXICAN DINNER

For Mandy's graduation, her family and friends went out to dinner at a Mexican restaurant. The restaurant served a bowl of chips for every three people, a bowl of salsa for every two people, and a bowl of guacamole for every four people. There were seven more bowls than people. How many people were at the dinner?

Strategies: systematic lists, subproblems

Difficulty: 2

Answer: 84 people (There were 91 bowls.)

©2004 Key College Publishing, *Instructor Resources: Crossing the River with Dogs,* Johnson/Herr/Kysh

Both contestants on the game show *Math Masters* were excellent mathematicians. The host gave each of them a positive whole number and told them that the product of their numbers was 15, 20, 24, or 28. The first contestant to determine the other's number would be the winner. Neither contestant could determine the other's number immediately. Both contestants thought about it and made some notes. Finally, one contestant determined the other contestant's number. What was the loser's number?

Strategies: systematic lists, eliminating possibilities, acting it out

Difficulty: 4

Answer: 5

THE FACTOR GAME

Two people are playing a game. Each factor of the number 72 is written on one of a series of cards and placed on the game board. The players alternate turns, with each player choosing one of the factors of 72. When a player chooses a number, he gets to take that number and all the factors of that number off the board. For example, if player one started the game by choosing 18, he would get to take 18, 9, 6, 3, 2, and 1 off the board, since those are all the factors of 18. It would then be player two's turn. He would choose one of the remaining numbers, and take that number and all of its factors off the board. The players continue to alternate turns until only the number 72 is left. The player who has to take 72 loses the game. It is possible, by careful play, for player one to win the game. What number should player one choose first, and how should he then play the rest of the game?

Strategies: eliminating possibilities, acting it out, manipulatives, working backwards, guess-and-check, easier related problems

Difficulty: 4

Answer: Take 6 first, and then adjust your next move: If player two takes 8 or 12, then you take 9. If player two takes 4, then you take 18. If player two takes 9, then you take 8. If player two takes 18, then you take 4. If player two takes 24, then you take 36. If player two takes 36, then you take 24.

©2004 Key College Publishing, *Instructor Resources: Crossing the River with Dogs*, Johnson/Herr/Kysh

Four people each hold a piece of paper with a positive whole number written on it. Each number is different. The sum of the four numbers is prime. Each pair of numbers has a greatest common factor larger than 1. Each of these greatest common factors is different. What is the smallest set of numbers that fit these conditions? (A greatest common factor is exactly what is sounds like, the largest number that is a factor of two particular numbers. For example, the numbers 8 and 32 have a greatest common factor of 8. The numbers 114 and 116 have a greatest common factor of 2.)

Strategies: eliminating possibilities, systematic lists, patterns, guess-and-check

Difficulty: 4

Answer: 6, 10, 15, 30

■ BLACK OR WHITE

Five logicians got together one evening to play some logic games. Their first game consisted of four players and one emcee who ran the game. The four players were Ryan, Torrey, Michael, and Bonnie. Janet acted as the emcee. Janet told the players that she was shuffling four white hats and three black hats and was going to put a hat on each of their heads. Each person would be given a chance to tell what color hat he or she was wearing. Janet told everyone to close their eyes. She then put a hat on each head and lined people up in single file, facing in the same direction. Ryan, the person in the back, could see all the other heads when he opened his eyes. Janet asked him if he knew what color hat he was wearing. He didn't know and said so. Torrey was next in line. She could not see Ryan, but could see the other two players. She also said she didn't know her hat color. Michael could see only Bonnie, and he said that he didn't know either. Bonnie, who was in front and could see no one else, knew the color of the hat on her head and announced it correctly. What color was her hat, and how did she know? (*Note:* Everyone can hear what the others said.)

Strategies: systematic lists, eliminating possibilities

Difficulty: 3

Answer: Bonnie's hat is white.

Martin, the school gardener, was an excellent logician. He gave a test to three young friends who were also excellent logicians: Ian, Stewart, and Paula. Martin told the three logicians, "I am going to blindfold each of you and then place a hat on each of your heads. The color of each hat is black, white, or green. It is possible for all the hats to be the same color, or for two hats to be the same color and one a different color, or for all three hats to be different colors. When I place the hats on your heads, I will remove your blindfolds at the same time. At that point, you will be able to see the hats on the heads of the other two people, but you won't be able to see your own hat. If you see a green hat, raise your right hand. Your task is to figure out the color of the hat on your own head."

So the logicians were blindfolded and the hats were placed. When the blindfolds were removed, all three logicians raised their hands. Stewart and Paula each immediately stated that they could name the color of the hat on their own heads. On hearing this, Ian said, "I don't know the color of the hat on my head, but I know what color my hat is not." What color was the hat on Stewart's head, and what color was the hat on Paula's head? Ian's hat was *not* what color? Explain all of your reasoning.

Strategies: systematic lists, eliminating possibilities, acting it out

Difficulty: 3

Answer: Stewart and Paula each wore a green hat. Ian's hat was not green.

OLIVER

Jill, Connor, David, and Lora performed in the musical *Oliver* last Friday, Saturday, and Sunday nights. A total of 750 people attended the three performances. More people attended each night than had attended the previous night, but the difference in attendance from one night to the next was not more than 10. Between the four of them, Jill, Connor, David, and Lora knew exactly one-third of Friday night's audience, exactly one-fourth of Saturday night's audience, and exactly one-fifth of Sunday night's audience. Out of the 750 people who attended the three performances, how many people did the four of them know? What was the attendance figure for each night of the performance?

Strategies: guess-and-check, eliminating possibilities

Difficulty: 3

Answers: The actors knew 195 people. The attendance figures were 243, 252, and 255.

The local community theater recently gave a four-night performance of *The Music Man*. The musical was performed in a large hall, where Connie set up chairs for each performance. She hadn't known exactly how many chairs to set up for each performance or in what arrangement, so she did it differently each night.

On the first night, Connie arranged the chairs so that there were four in each row. That night the house was sold out—every chair was occupied. On the second night, she added one chair to the total chairs from the previous night, then arranged the chairs so that there were five chairs per row. Again the house was sold out. On the third night, she again added one more chair to the total chairs from the previous night, then set up six chairs per row. Another sold-out house. On the fourth night, she again added one more chair to the total chairs from the previous night, then set up seven chairs per row. And again, the house was sold out.

Assuming that more than four people came to the first night's performance, find the next smallest possible number of people who attended each night of the performance (that is, the total number of chairs set up each night). Then answer this question: What was the total attendance for the four performances of the musical?

Strategies: guess-and-check, systematic lists, patterns, eliminating possibilities

Difficulty: 3

Answer: The numbers are 424, 425, 426, and 427. The total is 1702.

Suppose that a new version of Ping-Pong is invented and that the rules of the game are different. The game is still played between two people, and it takes 21 points to win the game. However, each time the score reaches a ratio that can be reduced, it is immediately reduced to lowest terms. For example, if the score reaches 4–2, it is reduced to 2–1. If it reaches 12–9, it is reduced to 4–3. If it reaches 14–12, it is reduced to 7–6, and so on. Even a score of 21–7 will be reduced to 3–1. What are the possible final scores for this game?

Strategies: acting it out, systematic lists, patterns, eliminating possibilities, working backwards

Difficulty: 3

Answer: 21 to 0, 1, 11, 13, or 17

VOLLEYBALL

Suppose that a new version of volleyball is invented and that the rules of the game are different. The game is still played between two teams, and it takes 15 points to win the game. However, each time the score reaches a ratio that can be reduced, it is immediately reduced to lowest terms. For example, if the score reaches 4–2, then it is reduced to 2–1. If it reaches 12–9, it is reduced to 4–3. If it reaches 14–12, it is reduced to 7–6, and so on. Even a score of 15–5 will be reduced to 3–1. What are the possible final scores for this game?

Strategies: acting it out, systematic lists, patterns, eliminating possibilities, working backwards

Difficulty: 3

Answer: 15 to 0, 1, or 11

MARCHING BAND

Al, Bart, Clark, Daryl, Eileen, Frosh, Greg, Henry, Irma, Jan, Kelly, Louis, Marcella, Nina, Olga, Pete, Quarrier, Russ, Shelley, Todd, Ulysses, Victoria, Wayne, Xue, and Yentl were all members of the River High School Marching Band. Their band director, Mr. Ryan, wanted them to practice a particular maneuver. They began in a five-by-five square, as shown below left. They then executed the maneuver, which left them in the positions shown below center. They then executed the maneuver again, which left them in the position shown below right. They executed this maneuver over and over again, and eventually they found that they were all in their original positions. How many times did they execute the maneuver?

START					AFTER 1 MANEUVER					AFTER 2 MANEUVERS				
A	B	C	D	E	K	E	G	B	O	P	O	F	E	S
F	G	H	I	J	L	F	A	C	J	H	L	K	G	J
K	L	M	N	O	P	H	I	N	S	Q	A	C	N	D
P	Q	R	S	T	Q	R	M	D	T	R	M	I	B	T
U	V	W	X	Y	Y	X	V	W	U	U	W	X	V	Y

Strategies: subproblems, diagrams, manipulatives, patterns, changing the representation

Difficulty: 4

Answer: 60 times

©2004 Key College Publishing, *Instructor Resources: Crossing the River with Dogs*, Johnson/Herr/Kysh

A group of 24 friends waited in line to see the opening of the latest sci-fi action flick. They had lined up in single file well before the show was to start. Because they'd arrived so early, they were restless and wanted to move around a bit. So, after waiting awhile, the 8th person in line moved up to the front of the line. The 16th person in line moved up directly behind that person (so was now 2nd in line), and the 24th person in line moved up directly behind those two people (and so was now 3rd in line). All 21 other friends moved back (so the original 1st person was now 4th, the original 2nd person was now 5th, and so on). They waited in line for a while longer and then switched places in line again: The person who was now 8th in line moved up to be 1st, the person who was now 16th in line moved up to be 2nd, and the person who was now 24th in line moved up to be 3rd, and everyone else moved back. After a while they did it again (8th to 1st, 16th to 2nd, and 24th to 3rd), and they continued this game over and over again until, eventually, they all found themselves back in their original positions. How many times did they switch places before they all found themselves back in their original positions?

Strategies: subproblems, diagrams, manipulatives, patterns, changing the representation

Difficulty: 4

Answer: 70 times

TRIKES FOR TIKES

Two 3-year-olds, Bob and Ray, are riding their trikes back and forth across a playground. Both of them ride at a constant speed (although their speeds are not equal), and they each take no time to turn around at either end of the playground. Bob starts at the west end of the playground, and Ray starts at the east end. They start at the same time and ride toward each other. They meet and pass each other 30 feet from the east end of the playground. When they reach the opposite end of the playground, they turn around and ride back toward each other. They meet again 14 feet from the west end of the playground. What is the length of the playground?

Strategies: diagrams, working backwards, algebra

Difficulty: 4

Answer: 76 feet

CROSSING THE STREET

I walk 6 feet/second, and I run three times as fast. Last night I was walking across a four-lane street. Halfway across the street, I noticed a car to my right that was just passing a fire hydrant. I started running, because the car was driving too fast for my sense of safety. It took 8 seconds for me to cross the street. One second after I reached the curb, the car sped by me. I then turned right and started walking again. I walked a distance that was three times the distance across the street until I reached the fire hydrant. How fast was the car moving in miles per hour?

Strategies: diagrams, unit analysis, subproblems, algebra

Difficulty: 3

Answer: 49 miles per hour

JOGGING ON THE FOOTBALL FIELD

Mike and Troy were jogging on the football field, which is 100 yards long and 50 yards wide. They started at opposite ends of the field, in the corners on the same side of the field. They ran toward each other. Each boy ran at a constant speed, although Mike ran faster than Troy. Mike passed Troy after Mike had run 60 yards and Troy had run 40 yards. They each continued on and ran all the way around the field, passing each other again at some point. It took Mike 15 seconds to finish his lap after he passed Troy the second time. How long did it take Troy to run around the field?

Strategies: diagrams, patterns, unit analysis, possibly acting it out

Difficulty: 2

Answer: 112.5 seconds

LONG BLOCK

I was taking a walk one day, when I noticed two people walking toward me. When they were passing a fire hydrant up ahead, I started counting seconds. From that point, it took me 15 seconds to pass them and another 12 seconds to reach the fire hydrant. Ten minutes after I passed them the first time, I passed them again. I realized that we were walking around the block in opposite directions. I know that I walk one yard per second. What is the distance all the way around the block?

Strategies: unit analysis, diagrams

Difficulty: 2

Answer: 1080 yards

©2004 Key College Publishing, *Instructor Resources: Crossing the River with Dogs*, Johnson/Herr/Kysh

A long, straight road runs through my neighborhood, and many people like to walk it. The road runs 1½ miles from the neighborhood to a stoplight and 1½ miles back to the neighborhood, a 3 mile round trip altogether. One day I was walking to the stoplight, at 3 mi/hr, when a runner passed me. He was also on his way to the stoplight. Six minutes later he passed me again on his way back to the neighborhood (he had turned around at the stoplight). I reached the stoplight 4½ minutes later and also turned around to head back to the neighborhood. By how much time did the runner beat me back to the neighborhood? Give your answer precisely, in minutes and seconds: for example, 7 minutes 56 seconds.

Strategies: unit analysis, subproblems, diagrams

Difficulty: 3

Answer: 24 minutes 18 seconds

WALKING AND RUNNING

On Saturdays I take a walk with my son. We walk down a long, straight road until we reach a stoplight, then we turn around and walk back home. On this particular Saturday, after we had been walking for a while, my son said he wanted to run. He ran all the way to the stoplight and then ran back to me. I was walking steadily toward the stoplight the whole time. When he got back to me, he turned around again and this time jogged all the way back to the stoplight again and then jogged back to me again. This time when he reached me, he turned around and walked with me to the stoplight (at my pace). My son's first round-trip (running) to the stoplight and back took 9 minutes. In other words, he met me again 9 minutes after he had left me. My son's second round-trip (jogging) to the stoplight and back took 6 minutes. Our final joint trip (one way—walking) to the stoplight took 1 minute. I walked at a constant 4.5 mi/hr the entire time. When my son was running, he ran at a constant rate the whole time. When he was jogging, he jogged at a constant rate the whole time. How fast did my son run, and how fast did he jog?

Strategies: diagrams, unit analysis, subproblems

Difficulty: 3

Answer: He ran at 11.5 miles per hour and jogged at 6 miles per hour.

I like to take walks with my son. We take this really long walk down a straight road until we reach a stoplight, then we turn around and walk back home. We took a walk last Sunday. After we had been walking for a while, my son said he wanted to run. He left me and ran all the way to the stoplight and then ran back to me. I was walking steadily toward the stoplight the whole time. When he got back to me, he turned around again and ran all the way back to the stoplight again and then ran back to me again. This time when he reached me, he turned around and walked with me to the stoplight (at my pace). My walking pace is a constant 4 miles per hour. My son's running pace is a constant 9 mi/hr. My son's first round-trip to the stoplight and back took 10 minutes. In other words, he met me again 10 minutes after he had left me. How long did my son's second round-trip take? Give your answer in minutes and seconds to the nearest second: for example, 12 minutes 43 seconds.

Strategies: unit analysis, diagrams, guess-and-check, algebra

Difficulty: 4

Answer: 3 minutes 51 seconds

In May 1995, *The Register-Guard* newspaper in Eugene, Oregon, reported on a group of fifth graders at Kenwood Elementary School in Bend, Oregon, and an interesting physics test they performed. The students marked off half a block in the school zone near their school, and they timed the cars and trucks that passed by. If it took less than 3.24 seconds for a vehicle to travel the marked distance, the students held up a sign that told the driver he or she was speeding. The speed limit in Bend school zones is 20 miles per hour when children are present, but the students gave motorists some leeway, holding up their sign only when speeds were clocked at 30 miles per hour or greater. What was the distance of the half block that the fifth graders marked off? Express your answer in a reasonable unit.

Strategies: diagrams, unit analysis

Difficulty: 1

Answer: 142.56 feet

©2004 Key College Publishing, *Instructor Resources: Crossing the River with Dogs*, Johnson/Herr/Kysh

Stephanie is the star of her softball team. Last week she hurt her knee a bit as she slid into home, so she had taped her knee for yesterday's game. Her knee was a little stiff, so she could run only 5 mi/hr yesterday, which is slower than her usual running speed. Midway through the game, she was at second base, with a 5-foot lead off of the base. The distance from second base to third base measured 60 feet. The pitcher pitched the ball, which traveled at 60 mi/hr. The distance from the pitcher's mound to home plate was 45 feet. Stephanie started running to third base the moment the pitcher pitched the ball. When the ball reached home plate, the batter hit the ball and it traveled at 75 mi/hr, directly to the right fielder, who was 180 feet away from home plate. The right fielder dropped the ball, then picked it up and threw it to third base. It took 2.5 seconds for the right fielder to drop the ball, pick it up, and release the throw. She threw the ball at 45 mi/hr to third base, which was 170 feet away. The third baseman caught the ball and tagged Stephanie. The act of catching the ball and applying the tag took 0.5 second. Was Stephanie safe or out?

Strategies: diagrams, unit analysis, subproblems

Difficulty: 2

Answer: She was safe by 0.2 second. She got to third in 7.5 seconds, and she was tagged at 7.7 seconds.

■ METRIC CLOCK

Time can often be expressed as hours : minutes : seconds. For example, the time 5:40:33 a.m. refers to 40 minutes 33 seconds after 5 o'clock in the morning. Several hundred years ago, there was a proposal to change the measurement of time to the metric system. The proposal was to divide each normal 24-hour day into ten metric "hours." Each "hour" would then be divided into 100 metric "minutes," and each "minute" would then be divided into 100 metric "seconds." There would no longer be the need for a.m. and p.m. Midnight would be 0:00:00 (or maybe 10:00:00) and noon would be 5:00:00 expressed in metric hours : minutes : seconds. This final exam started at 9:30 a.m. and ends at 11:30 a.m. Express these two times in metric hours : minutes : seconds.

Strategies: unit analysis, subproblems

Difficulty: 3

Answer: 9:30 = 3:95:83   11:30 = 4:79:17

The dashboard lights on my car were not working, so I could not see my speedometer. I could see my clock, but it showed only hours and minutes—no seconds—so I never knew how long it had shown the time indicated. However, I was able to use the freeway signs to approximate my speed. When I got to a sign that indicated how far I was from the next exit, I looked at the clock as I passed the sign and then again when I got to the exit. The results of this process are shown in the chart below. Assume I was traveling at a constant speed the entire time. What is the slowest speed I could have been traveling, and what is the fastest speed I could have been traveling? Answer in miles per hour.

Distance from sign to exit	Time when passing sign	Time when reaching exit
$1\frac{1}{4}$ mile	8:42	8:43
$1\frac{1}{2}$ mile	8:44	8:45
2 miles	8:46	8:48
$3\frac{1}{2}$ miles	8:49	8:54

Strategies: unit analysis, systematic lists, eliminating possibilities, organizing information

Difficulty: 4

Answer: The slowest speed is 45.4 miles per hour, and the fastest speed is 52.3 miles per hour.

RICKEY HENDERSON

On Tuesday, April 24, 2001, Rickey Henderson of the San Diego Padres walked. This walk tied Babe Ruth for the most walks in a career. On Wednesday, April 25, 2001, Henderson walked again and thus set the record for the most walks in a career. The distance from home plate to first base is 90 feet. When Henderson tied Ruth's record, he had walked the equivalent of 35.1 miles, rounded to the nearest tenth of a mile. The next day, when he broke the record, he had walked the equivalent of 35.2 miles, rounded to the nearest tenth of a mile. How many walks did Babe Ruth have in his career? (*Note:* There are 5280 feet in 1 mile.)

Strategies: unit analysis, guess-and-check

Difficulty: 3

Answer: 2062 walks

©2004 Key College Publishing, *Instructor Resources: Crossing the River with Dogs*, Johnson/Herr/Kysh

On the planet Rigel VII, time is measured in units of minutes, hours, days, weeks, months, and years. Because Rigel is in a different solar system, the length of time for each unit is different than it is on Earth. The following information refers to the time units on Rigel VII: There are twice as many weeks in a month as there are days in a week. There are as many days in a month as there are months in a year. There are half again as many hours in a day as there are minutes in an hour. There are 11 times as many hours in a day as there are days in a week. There are 235,224 minutes in a year. How many hours are in a week?

Strategies: unit analysis, subproblems

Difficulty: 4

Answer: 99 hours in a week

The inhabitants of planet Vega IV don't measure distances in inches, meters, feet, and so on. Their basic units of length are niche, toof, remet, dray, slime, and koli. Here are some of the relationships between these units:

There are three times as many toofs in a niche as there are remets in a toof.

There are half as many drays in a niche as there are remets in a slime.

There are as many drays in a toof as there are slimes in a koli.

There are 42 times as many drays in a niche as there are slimes in a koli.

There are 567 niches in a koli.

How many drays are in a slime?

Strategies: unit analysis, subproblems

Difficulty: 4

Answer: 23,814 drays in a slime

Kim walks her daughter Whitney to school every day. Kim's friend Denny also walks his daughter Brooke to school every day. Kim and Denny live next door to each other. It takes each of them x minutes to get to school (and x minutes to get back home) because they walk at the same speed. Normally they leave at the same time and walk together, but today Kim was in a rush and left her house y minutes before Denny left his. Kim got to school, dropped off Whitney, and walked back home. At some point along the way, she met Denny walking Brooke to school. At what point did they meet? (State your answer in terms of x and y, and clearly state whether your answer is from school to home or from home to school.)

Strategies: easier related problems, diagrams, acting it out, algebra

Difficulty: 4

Answer: $^{0.5y}/_x$ of the way from school to home

■ **PIE-EATING CONTEST**

Wesley and Gordon entered a pie-eating contest. Wesley had won the contest last year, so he started with a handicap: He had to start with three more pies than Gordon did. However, to partially offset this handicap, he was allowed to start eating 10 minutes earlier. Assume each boy's pie-eating rate remains the same throughout the contest (they don't necessarily eat at the same rate). Wesley started eating at 12:30. Gordon started eating at 12:40. They each had the same number of uneaten pies left at 1:00. Gordon finished eating his pies at 1:15. Wesley finished eating his pies at 1:20. How many pies did each boy have left at 1:05?

Strategies: subproblems, algebra, diagrams

Difficulty: 4

Answer: At 1:05, Wesley had 13½ pies left and Gordon had 12 pies left.

©2004 Key College Publishing, *Instructor Resources: Crossing the River with Dogs,* Johnson/Herr/Kysh

A classroom had two clocks. One worked normally, and the other one was broken, but still running in a peculiar way. For example, when the good clock read 2:00, the broken clock read 5:20. The broken clock worked as follows:

The second hand started at the 12 and went twice as fast as normal until it reached the minute hand, which is currently resting on the 4. When the second hand reached the minute hand, it paused for some number of seconds. After the pause, the second hand moved at normal speed and dragged the minute hand with it. This continued for 10 seconds, so during this time, the minute hand was dragged from the 4 to the 6. When the minute and second hand each reached the 6, the second hand left the minute hand behind and again went twice as fast as normal. The second hand continued at this speed until it met the minute hand again, which would now be resting on the 6. Again, the second hand paused for some number of seconds and then, at normal speed, dragged the minute hand with it for 10 seconds, from the 6 to the 8. Then, again, the second hand went twice as fast as normal until it reached the minute hand again. This pattern continued: twice as fast, pause, normal speed for 10 seconds as the minute hand was dragged, and so on. When the good clock reached 3:00, the broken clock again read 5:20. How many seconds did each pause last?

Important notes: The minute hand moves only when the second hand drags it. The hour hand moves in conjunction with the minute hand. Neither the hour hand nor the minute hand moves during the time that the second hand travels twice as fast as normal or as it pauses.

Strategies: systematic lists, subproblems, guess-and-check, diagrams

Difficulty: 4

Answer: 10 seconds

It was club day at Gladsville High School, and each club had set up a booth in the campus quad. There were four subject-area clubs (one was the Science Club), and their booths were set up next to one another. Each club was giving away some small item to interested people (one item was a button). Each booth was staffed by one person (one was named Dahlia). Determine where each club was located, the name of the person representing the club, and the item the club was giving away.

1. Holly's booth was directly between the booth giving away pencils (on the left) and the Music Club booth (on the right).

2. Tim's booth was directly between the Math Club booth (on the left) and the booth giving away erasers (on the right).

3. The four booths included the second booth from the left, the booth giving away stickers, Fred's booth, and the French Club booth.

4. Fred does not represent either the Math Club or the Science Club.

Strategies: eliminating possibilities, matrix logic, manipulatives

Difficulty: 1

Answer:

Math	Science	French	Music
Dahlia	Tim	Holly	Fred
stickers	pencils	erasers	buttons

Write a number, from 1 to 9, in each box so that each row, each column, and each small nine-box square contains each number from 1 to 9 exactly once. In other words, no number may appear twice in any row, column, or nine-box square.

	6	2			5	1		8
		8	7			2	6	
1			6	2			9	
	1	4			6	5		
6				8			3	2
2			5		7	6		
		7		5	1			6
4	2		8		3	9	5	
	8			7		3		4

Strategies: eliminating possibilities, manipulatives, guess-and-check, subproblems

Difficulty: 2

Answer:

7	6	2	3	9	5	1	4	8
9	5	8	7	1	4	2	6	3
1	4	3	6	2	8	7	9	5
8	1	4	2	3	6	5	7	9
6	7	5	1	8	9	4	3	2
2	3	9	5	4	7	6	8	1
3	9	7	4	5	1	8	2	6
4	2	1	8	6	3	9	5	7
5	8	6	9	7	2	3	1	4

Place a number, from 1 to 9, in each box so that each row, each column, and each small nine-box square contains each number from 1 to 9 exactly once. In other words, no number may appear twice in any row, column, or nine-box square.

5		4		8			9	
	6	1	4		9		3	5
7			2	6		8		
		7	8		6			3
3	4			9				7
		6	7			5	2	
	1			3	7			2
	7	3			8	9		
9		2	6			3	7	

Strategies: eliminating possibilities, manipulatives, guess-and-check, subproblems

Difficulty: 2

Answer:

5	2	4	3	8	1	7	9	6
8	6	1	4	7	9	2	3	5
7	3	9	2	6	5	8	1	4
2	9	7	8	5	6	1	4	3
3	4	5	1	9	2	6	8	7
1	8	6	7	4	3	5	2	9
6	1	8	9	3	7	4	5	2
4	7	3	5	2	8	9	6	1
9	5	2	6	1	4	3	7	8

Answers to Problem Sets

Introduction

1. **SOCCER GAME**

   There are many correct answers to this problem, depending on interpretation.

   If you use this problem as a Problem of the Day, do not get involved in which answer is "right." Let everyone present what they did.

   176 fives if every five is counted once.

   231 fives if the winners and losers exchange one five, but the winners and winners exchange two fives.

   352 fives if all fives are counted twice.

   Other answers may be possible as well.

2. **ELEVATOR**

   6 adults

3. **THEATER GROUP**

   18 men, 26 women

4. **DUCKS AND COWS**

   8 ducks and 4 cows

5. **STRANGE NUMBER**

   Any two-digit number where the sum of the digits is 11. For example: 92, 83, 74, 65, 56, 47, 38, 29.

Chapter 1, Problem Set A

1. **WORM JOURNEY**

   10 days

2. **UPS AND DOWNS OF SHOPPING**

   13 floors

3. **FOLLOW THE BOUNCING BALL**

   460 feet

4. **FLOOR TILES**

   320 tiles

5. **STONE NECKLACE**

   22 stones

6. **DANGEROUS MANEUVERS**

   Feline to Thirty-Nine to Bovine, 15 miles

   Lupine to Feline to Thirty-Nine, 17 miles

   Canine to Lupine to Feline, 14 miles

   Arachnid to Canine to Lupine, 18 miles; or Arachnid to Thirty-Nine to Feline to Lupine, 18 miles

   Canine to Arachnid to Thirty-Nine, 16 miles; Canine to Bovine to Thirty-Nine, 16 miles

   Lupine to Canine to Bovine, 10 miles

   Arachnid to Thirty-Nine to Feline, 7 miles

7. **RACE**

   Sabrina, Alma (3 meters behind), Ruby (3 meters behind), Lani (2 meters behind), Isabel (2 meters behind), and Becky (2 meters behind)

8. **A WHOLE LOTTA SHAKIN' GOIN' ON!**

   15 handshakes

9. **HAYWIRE**

   Cherlondia to Shirley to Darlene

   Carla to Cherlondia to Al to Max

   Sylvia to Henry to Carla to Cherlondia to Shirley to Darlene to Wolfgang

   Henry to Carla to Cherlondia to Shirley to Darlene to Wolfgang

   Shirley to Darlene to Sylvia to Henry

   Max to Henry to Carla to Sylvia to Dalamatia

   Max to Henry to Carla to Cherlondia to Dalamatia

   Cherlondia to Shirley to Darlene to Sylvia

   Messages can't be routed from Dalamatia to Henry.

10. ROCK CLIMBING
17 min 10 sec

11. CIRCULAR TABLE
17 inches

12. THE HUNGRY BOOKWORM
$2^5/_8$ inches

13. BUSING TABLES
7 new tables; $6^1/_5$ tables unbused

CHAPTER 1, CLASSIC PROBLEMS

15. THE WEIGHT OF A BRICK
3 pounds

16. THE MOTORCYCLIST AND THE HORSEMAN
40 minutes early

Chapter 1, Problem Set A, Version 2

1. THE GIFT EXCHANGE
280 gifts

2. AIR WARBUCKS
15 routes

3. COOKIE EXCHANGE
864 cookies

4. THE FINANCE COMMITTEE
36 calls

5. THE MERRY-GO-ROUND
16 seconds

6. CREEPING UP IN AGE
Allen, Marcus, Cindy, Lydia

7. CARISSA'S TABLE
1856 square inches

8. R-DOUBLE-7 RANCH
12 ostriches

9. MALFUNCTIONING ROTOTILLER
19 minutes

10. RUNDOWN
32 steps

11. THE DOG WHO LIKES TO RUN
6975 feet

12. TRAVELING ART SHOW
30 cities

13. GEORGETOWN RACE
Duesenberg–8 sec–Model T–4 sec–
Edsel–3 sec–Studebaker–7 sec–
Pierce-Arrow

14. CLASSROOM CEILING
144 panels

15. ARKANSAS CITIES
From Malvern to Pine Bluff: 50 miles by
way of Sheridan

From Camden to Pine Bluff: 72 miles by
way of Fordyce

From Benton to Arkadelphia: 72 miles by
way of Malvern and Sheridan

From Fordyce to Malvern: 95 miles by
way of Pine Bluff and Sheridan
(remember, Fordyce to Sheridan is closed)

From Benton to Camden: 122 miles by
way of Malvern, Sheridan, Arkadelphia,
and Gurden

16. WILDERNESS CAMP

They can't get a message to Eugenia because Carl has found only Glenda's camp and she has not found anyone else's. Hank can send a message to the leaders (by way of Carl) by sending it to Adrienne, to Freda, then to Carl.

a. You can't.

b. Route it through DuJannie and Bart to Glenda or Freda and Carl to Glenda.

c. You can't.

d. Route it to Bart, Eugenia, Hank, Adrienne to Freda, or Hank and Adrienne to Freda.

e. Send it through DuJannie, Hank, Adrienne, and Freda to Carl, or send it through Eugenia, Hank, Adrienne, and Freda to Carl.

f. You can't.

17. VACUUM CLEANER

44 seconds

18. CAMPAIGNING

a. 18 yards long

b. 10 yards between Tadashi and James

Chapter 2, Problem Set A

1. CARDS AND COMICS

CARDS @ $1.20	COMICS @ $0.60
5	0
4	2
3	4
2	6
1	8
0	10

2. TENNIS TOURNAMENT

There are 15 matches:

Justin v. Julie	Jamie v. Matt
Justin v. Jamie	Jamie v. Ryan
Justin v. Matt	Jamie v. Roland
Justin v. Ryan	
Justin v. Roland	Matt v. Ryan
	Matt v. Roland
Julie v. Jamie	
Julie v. Matt	Ryan v. Roland
Julie v. Ryan	
Julie v. Roland	

3. FREE CONCERT TICKETS

A = Alexis, B = Blake, and so on

ABCD	BACD	CABD	DABC
ABDC	BADC	CADB	DACB
ACBD	BCAD	CBAD	DBAC
ACDB	BCDA	CBDA	DBCA
ADBC	BDAC	CDAB	DCAB
ADCB	BDCA	CDBA	DCBA

4. APARTMENT HUNTING

In the ninth month, Plan A costs more ($640 per month versus $620 per month for Plan B). But for the whole year, Plan B costs more ($6,990 total versus $6,780 for Plan A).

5. **STORAGE SHEDS**

There are 10 possibilities or 16 possibilities. Here are 10:

8 ft × 8 ft	10 ft × 10 ft	12 ft × 12 ft	15 ft × 15 ft
8 ft × 10 ft	10 ft × 12 ft	12 ft × 15 ft	
8 ft × 12 ft	10 ft × 15 ft		
8 ft × 15 ft			

Note: There are six more possibilities if you consider, for example, an 8 ft × 10 ft shed to be different from a 10 ft × 8 ft shed.

A class discussion is very interesting here. Don't commit yourself to either answer.

---

6. **MAKING CHANGE**			$5	$10	$15	TOTAL
10 ways			4	0	0	20
			2	1	0	20
7. FINISHED PRODUCT			1	0	1	20
FIRST NUMBER	SECOND NUMBER		0	2	0	20
4	90		5	0	0	25
5	72		3	1	0	25
6	60		2	0	1	25
8	45		1	2	0	25
9	40		0	1	1	25
10	36		6	0	0	30
12	30		4	1	0	30
15	24		3	0	1	30
18	20		2	2	0	30
			1	1	1	30
8. BASKETBALL			0	3	0	30
14 ways			0	0	2	30
			7	0	0	35
9. TWENTY-FOUR			5	1	0	35
15 ways			4	0	1	35
			3	2	0	35
10. TARGET PRACTICE			2	1	1	35
35 total scores			1	3	0	35
			1	0	2	35
11. TANYA'S TERRIFIC T-SHIRTS			0	2	1	35

11. TANYA'S TERRIFIC T-SHIRTS

There are 40 ways (or 41 if you count not spending any money).

$5	$10	$15	TOTAL
1	0	0	5
2	0	0	10
0	1	0	10
3	0	0	15
1	1	0	15
0	0	1	15

$5	$10	$15	TOTAL
8	0	0	40
6	1	0	40
5	0	1	40
4	2	0	40
3	1	1	40
2	3	0	40
2	0	2	40
1	2	1	40
0	4	0	40
0	1	2	40

13. ARCHERY PUZZLE
6 arrows: 17, 17, 17, 17, 16, 16

14. WHICH BARREL WAS LEFT?
The 19-gallon barrel

Chapter 2, Problem Set A, Version 2

Note: Some of the answers given for this problem set state only the number of items in a list. Student work should show complete lists.

1. RIDE TICKETS
29 ways

2. LEARNING THE HARD WAY
24 ways

3. RENTING A CAR
The second plan is cheaper up through day 7. He should probably choose that one. The two plans present equal cost on day 8, after which the first plan becomes cheaper.

4. WAYLON'S CANDY
21 ways

5. FENCING WITH NEIGHBORS
7 ways

6. ARCADE
The possible scores are 75, 80, 85, 90, 95, 100, 105, 110, 115, 120, 125, 130, 140, and 150 (all the multiples of 5 from 75 to 150 except 135 and 145).

7. RUDY'S SHOT TOTALS
10 ways

8. SKIING BIKERS
8 ways

9. SUBJECT-VERB-OBJECT
168 sentences

10. MAKING CHANGE FOR 70 CENTS
16 ways

11. MAKING CHANGE FOR 55 CENTS
11 ways

Chapter 3, Problem Set A

1. SQUARE ROOTS
66 and 92

2. HOW MANY LINES?
47 lines. Notice that no upper limit is given. The students should figure out what a reasonable upper limit is.

3. EGGS IN A BASKET
119 eggs

4. DARTBOARD
23, 58, 31, 6, 15

5. FIND THE NUMBER
$abcd$ = 2178 (2178 × 4 = 8712)

6. WOW, WOW, SO COOK!

	W	O	W
	7	5	7
	W	O	W
	7	5	7
+		S	O
		4	5
C	O	O	K
1	5	5	9

7. NELSON + CARSON = REWARD

```
  N E L S O N
  5 2 6 4 8 5
+ C A R S O N
  1 9 7 4 8 5
  -----------
  R E W A R D
  7 2 3 9 7 0
```

8. THE THREE SQUARES

Chris is 19, Phyllis is 23, and Bob is 28.

9. TO TELL THE TRUTH

If Dog 1 answers the question, then he must have said, "I am a truth teller," because that is the only possible answer to the question "Are you a truth teller or a liar?" Liars lie and truth tellers tell the truth, so everyone would say they were a truth teller. Therefore, Dog 2 is a truth teller and Dog 3 is a liar, but Dog 1 is undetermined.

However, suppose Dog 1 does not answer the question. Suppose he says, "The sky is blue." Now Dog 2 is a liar, because Dog 1 didn't say he was a truth teller. And Dog 3 could be a truth teller or a liar depending on whatever Dog 1 actually said and whether it was true or not, and depending on which, if any, dog(s) is named Joe.

10. FOUR COLLEGE ROOMMATES

	Height	Age	Weight
1	Thuy	Jerel	Nate
2	Miguel	Thuy	Jerel
3	Nate	Miguel	Thuy
4	Jerel	Nate	Miguel

11. THE LETTER FROM COLLEGE

```
  S E N D
  9 5 6 7
+ M O R E
  1 0 8 5
  -------
M O N E Y
1 0 6 5 2
```

He wanted $106.52.

CHAPTER 3, CLASSIC PROBLEMS

13. TURKEYS

Missing digits are 3 and 2. The total is $367.92; turkeys cost $5.11 each.

14. SCRAMBLED BOX TOPS

It takes one drawing. Draw one marble from the box labeled BW. If it is B, then there are two B's in BW, 2 W's in BB, and BW in WW. If it is W, there are two W's in BW, 2 B's in WW, and BW in BB.

15. THE TRIAL

Choose the door labeled "A tiger is in the other room."

16. THE CONSPIRATORS' CODE

```
    F L Y
    5 9 8
    F O R
    5 0 7
+   Y O U R
    8 0 4 7
    -------
    L I F E
    9 1 5 2
```

1. **SITTING IN THE PARK**
   14 or 74 people

2. **DANCING IN PE**
   61 students

3. **MARBLE ARRANGEMENTS**
   62 marbles
   In piles of 2 there would be none
   left over.

4. **STATE QUARTERS COLLECTION**
   68 quarters

5. **A CUBE ROOT**
   41

6. **BORED TEENAGERS**

   ```
       C O W S
   +   B A R N
   ─────────────
     T R E E S
   ```

   Versions 1 and 2:

   ```
       8 6 4 2
   +   7 3 5 0
   ─────────────
     1 5 9 9 2
   ```

   Version 3:

   ```
       8 2 6 4
   +   5 7 3 0
   ─────────────
     1 3 9 9 4
   ```

7. **ONE-HOUR CASSEROLE**
   There are two answers possible. One
   answer is shown. The other answer
   interchanges the digits U and N so that
   U = 4 and N = 2.

   ```
       H O U R
       6 0 2 9
   +   S A N D
       7 5 4 8
   ─────────────
     G L A S S
     1 3 5 7 7
   ```

8. **NAPA VALLEY TOUR**
   There are two answers possible. One
   answer is shown. The other answer
   interchanges the digits F and T so that
   F = 7 and T = 6.

   ```
       F R E E
       6 4 8 8
   +   T O U R
       7 0 9 4
   ─────────────
     W I N E S
     1 3 5 8 2
   ```

9. **SHUTOUT**
   One of four possible answers is shown
   below. C and L are interchangeable, and
   E and I are interchangeable.

   ```
       S O C C E R
       5 2 3 3 4 0
   +   G O A L I E
       7 2 9 8 6 4
   ─────────────────
     N O S C O R E
     1 2 5 3 2 0 4
   ```

10. **SECRET TO MONOPOLY**
    Very tough. This is the only solution.

    ```
        H O U S E S
        8 7 9 6 4 6
    +   H O T E L S
        8 7 3 4 2 6
    ─────────────────
      C O N T R O L
      1 7 5 3 0 7 2
    ```

11. FRUITS AND VEGETABLES CRISSCROSS

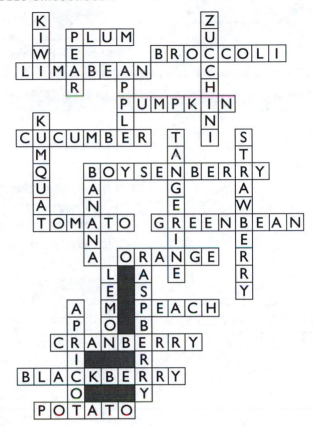

12. BASEBALL CRISSCROSS

Chapter 3, Problem Set B

I. THE SIDEWALK AROUND THE GARDEN
172 feet

2. A NUMBER OF OPTIONS
24 ways

3. GOOD DIRECTIONS?
Go two blocks east and one block south.

4. HIGH SCORERS
Haley, 25; Shauntae, 23; Martina, 19;
Donna, 17; Kellene, 11

5. WAYS TO SCORE
19 ways

Chapter 3, Problem Set B, Version 2

I. LOG CUTTING
12 minutes

2. WEIRD WORDS
30, or 36 if Y is included as a vowel

3. GETTING AROUND TOWN
Go 3 blocks east to 16th, turn left. Go 5
blocks north to I, turn left. Go 4 blocks
west to 12th, turn left. Go 2 blocks south
to K, and you are there. Or go 3 blocks
east to 16th, turn left. Go 2 blocks north
to L, turn left. Go 5 blocks west to 11th,
turn right. Go 2 blocks north to J, turn
right. Go 1 block east to 12th, turn right.
Go 1 block south to K, and you are there.

Both routes are 14 blocks long. The
first route has fewer turns (three instead
of five).

4. SLEEPY BABIES

```
    B  A  B  Y
    8  7  8  1
  + B  A  B  Y
    8  7  8  1
  _____
  Y  A  W  N  S
  1  7  5  6  2
```

5. VIDEO GAMES
14 ways

Chapter 3, Problem Set B, Version 3

I. THE BIKE PATH AROUND THE LAWN
212 feet

2. FAST FOOD
24 ways
hamburger, fries, coffee
hamburger, fries, milk
hamburger, fries, soda
hamburger, fries, milkshake
hamburger, onion rings, coffee
hamburger, onion rings, milk
hamburger, onion rings, soda
hamburger, onion rings, milkshake

chicken nuggets, fries, coffee
chicken nuggets, fries, milk
chicken nuggets, fries, soda
chicken nuggets, fries, milkshake
chicken nuggets, onion rings, coffee
chicken nuggets, onion rings, milk
chicken nuggets, onion rings, soda
chicken nuggets, onion rings, milkshake

hot dog, fries, coffee
hot dog, fries, milk
hot dog, fries, soda
hot dog, fries, milkshake
hot dog, onion rings, coffee
hot dog, onion rings, milk
hot dog, onion rings, soda
hot dog, onion rings, milkshake

3. A DARING CATFISH
9 days and 9 nights

4. FIND MY NUMBERS
13, 17, 19, 23, 25

5. FOOTBALL SCORES
25 ways

Chapter 3, Problem Set B, Version 4

1. **TRAILERS**
   24 trailers

2. **TRAILER PAD**
   $35\frac{1}{3}$ ft^3 or 61,056 in.^{3}

3. **KITE CHASING**
   8 steps to the right of original position

4. **FIELDING PRACTICE**
   22 ways

5. **SLEEPING IN THE TENT**

   ```
       T  E  N  T
       8  4  5  8
    +  S  I  T  E
       1  9  8  4
    ────────────────
    S  L  E  E  P
    1  0  4  4  2
   ```

Chapter 4, Problem Set A

1. **THE FISHING TRIP**
   Sally first, Mickey second, Woody third, Marta fourth

2. **DIVISION I ATHLETES**
   Russ Krieger
   Don Hicks
   Pamela Drake
   Stephanie Braun

3. **A DAY ON THE LAKE**
   John – water-ski
   Ellen – swim
   Armand – drive boat
   Judy – dive off rocks

4. **BROTHERS AND SISTERS**
   Scott – nail polish
   Lynn – computer
   Cheryl – shampoo
   Greg – panda

5. **STATE QUARTERS**
   Robin – Delaware
   Nicholas – Rhode Island
   Jen – Pennsylvania
   Miles – New York
   Dustin – Massachusetts

6. **CABINET MEMBERS**
   Georgianne – president
   Norma – vice president
   Inez – secretary of state
   Paula – secretary of education
   Colleen – secretary of treasury

7. **ECOLOGY EXPERTS**
   Abbie – water quality
   Bridget – soil contamination
   Cynthia – air pollution
   Dena – biological diversity

8. **VOLLEYBALL TEAM**
   Elaine – outside hitter, freshman
   Kelly – setter, junior
   Shannon – middle blocker, sophomore

9. **MUSIC PREFERENCES**
   Jack Mullin – country-western
   Mike Hardaway – rock
   Adele Higgins – jazz
   Edna Richmond – classical

10. **GRADUATE SCHOOL APPLICATIONS**
    Gladys Burbank – vet school
    Cathy Gunderson – seminary
    Alan Williams – med school
    Hank Collins – law school

11. **SUSPECTS**
    Connie Wilde – purple hair
    Morgan Theeves – scar
    Dana Steele – tall and blonde
    Cary Fleece – birthmark

12. **ANNIVERSARIES**
    Jorge married Lorna in May, 11 years ago.
    Ahmed married Tara in July, 12 years ago.
    Pierre married Nylia in June, 13 years ago.

13. PAYING THE BILLS

Rosita – apartment – scuba diving
Kathleen – car – rock climbing
Michelle – dorm – swimming
Ginny – clothes – snorkeling

14. CLASS SCHEDULES

a. Jillian's schedule
1 Band
2 PE
3 Science
4 Math
5 Lunch
6 English
7 History

b. Todd's schedule
1 PE
2 Math
3 Drama
4 Science
5 Lunch
6 English
7 Typing

c. Leanne's schedule is impossible because she has to take both Math and PE during second period and has no class to take during seventh period. Students can set up a schedule for her including Band or Drama but not both. They will then have to give her a seventh-period class—either History or Typing.

d. Mea's schedule has many possibilities.

e. Part 1: José's first schedule looks like this:
1 PE
2 Math
3 Drama
4 Science
5 Lunch
6 English
7 History

Part 2: Because first period PE is closed, there is no way for José's schedule to work. He won't be able to take Drama because he will have to take Science during third period to make his schedule work. So he must pick another elective, either Band or Typing.

Part 3: Now, with the new sixth-period science class, José's schedule will work. (*Note:* Math and PE can be switched. The schedule will also work with PE in second period, Math in fourth period, and Lunch in fifth period.)
1 English
2 Math
3 Drama
4 Lunch
5 PE
6 Science
7 History

CHAPTER 4, CLASSIC PROBLEM

16. THE ENGINEER'S NAME

The engineer's name is Smith.

Chapter 4, Problem Set A, Version 2

1. THE HOBBYISTS

Leisa, model airplanes; Elaine, model railroading; Consuelo, rocketry; Brittney, tropical fish

2. MIXED DOUBLES TENNIS

Mack Katricz and Sherry Blue Cloud versus Timothy Amaya and Diana Ostergard

3. SUMMER JOBS

LaTisha, lifeguard; Zack, clerk; Steve, food server; Michelle, construction worker

4. SPORTS ARE ALL RELATIVE

Bonnie and Michael, basketball; Ryan, tennis; Stefan, bowling

5. MATH DEPARTMENT MEETING

Fara, statistics; Laurie, finite; Elaine, algebra; Maile, calculus; Cliff, tech math

6. **NEXT YEAR AT COLLEGE**
Jenny Oslowski, Penn Valley; Akinte Reynoso, Saint Mary's; Chuck Penigar, University of Memphis; Norma Van Hee, Sierra Tech

7. **LOOMIS DAY PARADE**
First, Wayne, baton; second, Stacy, band; third, Mort, dance; fourth, Cloe, equestrian

8. **STUDENT ACHIEVEMENT AWARDS**
Matt Kinsella, math; Ginny Perata, computers; Clyde Macomber, English; Velma Sholseth, physics

9. **NOVEMBER ELECTIONS**
Marguerite Hardy, governor; Darcie Cusack, assembly; Brent Tomfohrde, senator; Amir Wyckoff, president

Chapter 4, Problem Set B

1. **PHONE NUMBER**
492-2804

2. **THE BILLBOARD**
7 lines

3. **A WORTHY SUITOR**
14 ways

4. **THOSE AMAZING NAMES**
There are two possible answers.

	E	L	I	S	A
	1	9	5	6	8
+ A	J	U	D	Y	
8	4	3	0	2	
E	D	U	A	R	D
1	0	3	8	7	0

	E	L	I	S	A
	1	4	5	6	8
+ A	J	U	D	Y	
8	9	3	0	2	
E	D	U	A	R	D
1	0	3	8	7	0

5. **THE SPORTING EVENTS**
Ed, baseball, Monday; Judy, frisbee, Tuesday; Mama, golf, Thursday; Lisa, soccer, Friday

Chapter 4, Problem Set B, Version 2

1. **PACIFIC RIM**
28 routes

2. **PRIME JACKS**
Most: Dawn 79 and Ellen 89 can combine for 168 jacks, which can be separated into same-sized piles in 16 ways.
Least: Betty 29 and Ellen 89 combine for 118 jacks, which can be separated into same-sized piles in only 4 ways.

3. **RELATIONSHIPS**
Jordan is Mela's grandson. Tina is Brandon's daughter-in-law.

4. **TO TELL THE TRUTH**
Abe. This is the only thing you can tell for sure.

5. **THE ACTION NEWS TEAM**
Alex, producer, 20 years; Chris, director, 10 years; Pat, anchor, 5 years; Sam, sports, 15 years

Chapter 4, Problem Set B, Version 3

1. **DOOR IN AND DOOR OUT**
30 ways

2. **JERRY'S AGE**
44 years old

3. **KLINGONS LIKE WORD ARITHMETIC TOO**

C	R	E	K	L	G
1	8	4	0	6	5

+ R	U	T	A	N	G
8	7	2	3	9	5

C	K	G	L	E	L	K
1	0	5	6	4	6	0

4. **THE SHADOW KNOWS**

He needs to walk 8 miles with his shadow in front of him and 2 miles with his shadow on his right.

5. **THE NEW FALL SEASON**

Wall Street Blues, Tuesday, variety; *All in Favor,* Thursday, news; *Murphy's Law,* Wednesday, drama; *Fifty-Fifty,* Monday, comedy

Chapter 4, Problem Set B, Version 4

1. **LUCKY SVEN**
30 prices

2. **LETTER PUZZLE**

0	1	2	3	4	5	6	7	8	9
M	A	G	Y	H	C	W	R	L	D

3. **TWO GUARDIANS**
Door two

4. **RELATIVES**

Half siblings are Tom and Phillip, Phillip and Jeannine, Jeannine and Nancy, Nancy and Woody. Step-siblings are Tom and Jeannine, Jeannine and Woody.

5. **WAYNE'S WORLD**

Dana Campbell, keyboard; Mike Algar, guitar; Garth Carvey, drums; Wayne Myers, bass

Chapter 5, Problem Set A

1. **SEQUENCE PATTERNS**
   a. 2, 5, 10, 17, **26, 37, 50**. Add next odd number. Or, each term is 1 more than a perfect square.
   b. 64, 32, 16, 8, 4, **2, 1, $^1/_2$**. Divide by 2. Or, each term is a descending power of 2.
   c. 5, 10, 9, 18, 17, 34, 33, **66, 65, 130**. Multiply by 2 and then subtract 1.
   d. 1, 3, 7, 13, 21, **31, 43, 57**. Add the next even number.
   e. 1, 5, 13, 29, 61, 125, **253, 509, 1021**. Starting from 4, the differences double each time. Or, add the next higher power of 2, starting from 4.
   f. 1, 5, 13, 26, 45, 71, **105, 148, 201**. Start with 4; the difference of the differences increases by 1.
   g. 1; 2; 6; 24; 120; 720; **5,040; 40,320; 362,880**. Multiply by the next greater number. Or, each term is *n*! (*n* factorial).

2. **MORE SEQUENCE PATTERNS**
   a. 243, 81, 27, 9, 3, **1, $^1/_3$, $^1/_9$, $^1/_{27}$**. Divide by 3.
   b. 4, 9, 8, 13, 12, **17, 16, 21, 20**. Add 5, then subtract 1.
   c. 4, 5, 8, 13, 20, 29, 40, **53, 68, 85, 104**. Add 1, 3, 5, 7, 9, ...
   d. 3, 7, 13, 21, 31, 43, **57, 73, 91, 111**. Add increasing multiples of 2.
   e. 5, 7, 11, 19, 35, 67, **131, 259, 515, 1027**. Add 2, 4, 8, 16, 32, ...
   f. 1; 4; 13; 40; 121; 364; **1,093; 3,280; 9,841; 29,524**. Add the next highest power of 3, or 3, 9, 27, 81, ...
   g. 3, 1, 4, 5, 9, 14, 23, 37, **60, 97, 157, 254**. Fibonacci sequence: Add the previous two terms.

h. 7, −3, 10, −13, 23, −36, 59, **−95, 154, −249, 403**. Fibonacci sequence: Subtract the previous two terms.

i. 1, 2, 3, 6, 11, 20, 37, **68, 125, 230, 423**. Similar to a Fibonacci sequence, but add the previous three terms.

3. **AIR SHOW**
   400 people. The pattern is the square of the number of rows.

4. **RECTANGULAR DOTS**
   $34 \times 35 = 1190$ dots

5. **PENTAGONAL NUMBERS**
   425 dots. Add a triangular number and a square number. The triangular number has one fewer dot per side than the square number does.

6. **BEACH BALL**
   Orange

7. **LAST DIGIT**
   2^{57} ends in 2. The pattern goes 2, 4, 8, 6, 2, 4, 8, 6, ...

8. **JUGGLING**
   The 100th catch is a yellow ball in the left hand; the blue ball would be in the air, on its way to the right hand. Gessop's assumption that there's a 50-50 chance the blue ball would end up in his right hand is incorrect, because the ball could be in any one of the three positions.

9. **EMAIL VIRUS**
   6,377,551 emails

10. **BEES**
    231 ancestors. This is a Fibonacci sequence added up.

11. **PASCAL'S TRIANGLE**
    1, 6, 15, 20, 15, 6, 1
    1, 7, 21, 35, 35, 21, 7, 1
    1, 8, 28, 56, 70, 56, 28, 8, 1
    1, 9, 36, 84, 126, 126, 84, 36, 9, 1

12. **OTHER PATTERNS IN PASCAL'S TRIANGLE**
    Answers will vary.

13. **COIN FLIPS**
    Three coins: HHH, HHT, HTH, HTT, THH, THT, TTH, TTT. With four coins there are 16 ways. The number of heads or the number of tails show up in Pascal's triangle. For example, with three coins, there is one way to get three heads, three ways to get two heads, three ways to get one head, and one way to get zero heads, for a total of eight outcomes. The 1, 3, 3, 1 row of Pascal's triangle describes these possible three-coin combinations.

14. **FREE INTERNET**
    89 minutes 50 seconds

CHAPTER 5, CLASSIC PROBLEMS

17. **THE HAT THAT DIDN'T SELL**
    $0.51

18. **TOWER OF HANOI**
    $2^{64} - 1$

Chapter 5, Problem Set A, Version 2

1. **SEQUENCE PATTERNS AGAIN**
   a. 5, 8, 11, 14, 17, **20, 23, 26**. Explanations may vary. Add 3 to get the next term.
   b. −1, 1, 3, 5, 7, **9, 11, 13**. Add 2 to get the next term.
   c. 2, 3, 5, 8, 12, **17, 23, 30**. Add 1, add 2, add 3, and so on.
   d. 3, 2, 5, 7, 12, **19, 31, 50**. Fibonacci sequence: Add two consecutive terms to get the next term.
   e. −1, 5, 4, 9, 13, **22, 35, 57**. Fibonacci sequence: Add two consecutive terms to get the next term.

(Answer continues on next page.)

f. 4, 5, 7, 10, 10, 15, 13, **20, 16, 25**.
   Double series: One is adding 5, the
   other is adding 3.

g. $-2, -1, 1, 4, 8,$ **13, 19, 26**. Add
   consecutive numbers to get the next
   term: add 1, then add 2, add 3, ...

h. 5, 7, 11, 19, 35, **67, 131, 259**. Add
   consecutive powers of 2.

2. **THE GREAT SALE**
   797,160 people, not including Asa

3. **MMM, MACARONI AND CHEESE**
   $1/3$ of the macaroni and cheese

4. **BIRD AND BEEF**
   2450 customers

5. **MAILING LISTS**
   305,175,780 catalogs

6. **THE ANTS COME MARCHING IN**
   55 ants

7. **PARTY TIME**
   6 weeks (728 parties)

8. **YOUR OWN SEQUENCES**
   Answers will vary. Students can set up
   the problems as in the text and exchange
   papers to check one another's sequences.

Chapter 5, Problem Set B

1. **LEGAL EAGLES**
   Ostrom, tan, first; Savidge, burgundy,
   fifth; Stetson, black, fourth; Neumann,
   blue, second; Schoorl, silver, third.
   Neumann is a man.

2. **RUDY'S CLOTHES RACK**
   $7/11$

3. **ROO AND TIGGER**
   Roo won by 4 feet.

4. **GOLF MATCH**
   Diana will tee off second.

5. **COMIC OF THE MONTH**
   $22.42

Chapter 5, Problem Set B, Version 2

1. **SOMETHING BUT THE TRUTH**
   Rick, innocent; Louise, accomplice;
   Manny, main burglar

2. **YALE RECORD CLUB**
   $6.99 for each CD, $3.87 for shipping,
   and 38 CDs cost $269.49.

3. **TEN POSTS**
   600 meters

4. **THE PET STORE CONTEST**
   $11/24$

5. **WORLD WIDE WIDGETS**
   Woody, Wichita, bookkeeper; Ned,
   Dodge, manager; Gus, Concordia, vice
   president; Dick, Lawrence, window
   washer; Jake, Belleville, mailroom clerk

Chapter 5, Problem Set B, Version 3

1. **BURBANK NEIGHBORS**
   Stanley, maroon, fourth; Neuerburg, red,
   second; Stump, green, first; Frick, yellow,
   fifth; Wahhab, blue, third

2. **BUCKS FOR CLUCKS**
   Choose egg 3. It's a winner if egg 3 has
   the true statement, and it has a 50%
   chance of being a winner if egg 1 has the
   true statement.

3. **KANGA AND ROO**
   Roo won by 8 feet.

4. **DOG AND TREES**
568 meters

5. **PRODUCT OF OUR TIMES**
16 times

Chapter 5, Problem Set B, Version 4

1. **NEW CAR OPTIONS**
53 styles

2. **SHEET CAKE**
Row 5, column 5

3. **PLANET THREA**
$^1/_5$

4. **CAMP SACRAMENTO**
Lisa Horlick, rings; Danny Horlick, slide; Jamie Walker, tire swing; Justin Walker, monkey bars; Jacob Bland, rock

5. **PLAYING DETECTIVE**
Joe turned off the light switch.

Chapter 6, Problem Set A

1. **DIMES AND QUARTERS**
8 quarters, 13 dimes

2. **MARKDOWN**
$34.30

3. **TAX**
The basketball cost $14.49, and the tax was $1.05.

4. **REFINANCING**
36 months

5. **COLLEGE TOWN TRANSIT TICKETS**
24 rides

6. **CHECKING ACCOUNT**
39 checks

7. **BASEBALL CARDS**
35 cards

8. **A BUNCH OF CHANGE**
14 dimes, 25 nickels, 19 quarters

9. **UNDERGRADS**
13,810 women

10. **STRANGE COINCIDENCE**
Ronnie, Apt. 31; Harry, Apt. 47

11. **GOING HOME FROM COLLEGE**
45 miles

12. **RIDING A HORSE**
3 miles per hour

13. **WOMEN'S WORLD CUP SOCCER**
17.5 yards

14. **TELEPHONE SOLICITOR**
65, 66, and 67 calls all work.

15. **EQUAL VOLUME**
6.75 inches

16. **FREE THROWS**
There are several possible answers. The most exact is 245. Several other answers also work because of rounding: 237, 238, 244, 246, 251, 252, and 258.

17. **CELL PHONE PLANS**
Marathon is less expensive for less than 165 minutes. Toucan is less expensive from 165 minutes up to 415 minutes. Marathon is less expensive after 415 minutes.

CHAPTER 6, CLASSIC PROBLEM

19. **THE FIVE NEWSBOYS**
The Jones brothers sold 220 more papers.

Chapter 6, Problem Set A, Version 2

1. **QUARTERS, DIMES, AND NICKELS**
   4 quarters, 12 dimes, and 16 nickels

2. **CHANGE**
   7 nickels, 10 dimes, and 13 quarters

3. **ROCKS**
   Christopher has 22 rocks, and Gordon has 9 rocks.

4. **LONG JOURNEY**
   75 miles. Answer to alternate question: $1\frac{1}{2}$ hours.

5. **HEALTH CLUB FEES**
   After 15 months, the two clubs will cost the same. From the 16th month on, the new club will be cheaper.

6. **T-SHIRTS**
   $7.80

7. **CENTRAL VIRGINIA COLLEGE**
   522 sophomores, 609 freshmen; so there are 87 more freshmen.

8. **MUTUAL FUNDS**
   $78,300

9. **COMPACT DISCS**
   11 compact discs

10. **STICKERS**
    Cici started with 18 stickers. Amatina started with 54.

11. **LOTS OF COINS**
    27 quarters, 31 dimes, and 22 nickels

12. **CHANGE IN THE DRAWER**
    13 quarters, 41 dimes, and 16 nickels

13. **BROADWAY BOUTIQUE**
    $18.70

14. **FITNESS COSTS MONEY**
    After 15 months, Being Fit is less expensive.

Chapter 6, Problem Set B

1. **DAILY ROUTINE**
   144 days

2. **AFTER THE FOOTBALL GAME**
   15 people

3. **CATS**
   201 cats

4. **STOCK MARKET**
   Nita McDonald, Ford, lost $300; Tina Kortright, IBM, made $200; Luann Edwards, Xerox, made $400; Vickie McElhatton, ATT, made $700

5. **LARRY LONGWAY**
   3, 3, and 8

Chapter 6, Problem Set B, Version 2

1. **SPARE CHANGE**
   14 ways

2. **SKI TRIP**
   $840

3. **EXPENSIVE MISSILE**
   $3,596,182,074

4. **FROG CHORUS**
   1908 seconds

5. **THE AYER FAMILY**
   Lucy, 14, August, Monday; Paul, 7, July, Wednesday; Lawrence, 11, January, Thursday; Don, 15, November, Saturday

Chapter 6, Problem Set B, Version 3

1. **READY FOR "THE SHOW"**
   $^3/_{11}$

2. **BIKE RACE**
   11 volunteers

3. **STORKE TOWER**
   21 minutes 12 seconds

4. **COLLEGE ROOMMATES**
   Glory Alder, black, chemistry; Helen Carlson, brown, physics; Joan Daniels, blonde, accounting; Irene Bonds, red, English

5. **LUNCH MONEY**
   He spent $2, $3, and $16.

Chapter 6, Problem Set B, Version 4

1. **GRANDPA'S ATTIC**
   21 grandchildren

2. **THE MATH TEST**
   1, 30; 2, 15; 3, 10; 5, 6

3. **FAMILY OUTINGS**
   Erin, Presidents' Day, amusement park, hot dogs; Andy, Memorial Day, zoo, pizza; Jenny, Fourth of July, museum, burgers; Neal, Labor Day, county fair, deli sandwiches

4. **RHO-SHAM-BO**
   Craig wins.

5. **POOR SCORE**

Hole	1	2	3	4	5	6	7	8	9	Total
Bob A	4	4	7	2	3	4	4	5	3	36
Bob B	3	4	4	4	5	3	3	4	6	36

Chapter 7, Problem Set A

1. **COFFEE**
   6 ounces

2. **SHARING EXPENSES**
   Many answers are possible for how to accomplish paying everyone back. The total amount spent was $69.40, so each person's share is $13.88. Leroy needs $6.12. Alex owes $8.88. Kulwinder owes $0.88. Maxx needs $2.12. Bobbi needs $1.52.

3. **AIRPLANE SEATS**
   240 seats

4. **SIX SQUARES**
   42 feet

5. **SHADED AREA**
   $100 - 25\pi$, or 21.46 square centimeters. *Note:* The $^1/_4$ circles are all supposed to be the same size.

6. **SAVINGS PLAN**
   12%

7. **TEST AVERAGE**
   94%

8. **CAR TRIP**
   45 minutes

9. **FARGO**
   60.3 miles per hour

10. **TEST TRACK**
    8 hours

11. **BOX**
    1620 in.^{3}

12. **SWEETENED CEREAL**
    34 ounces

13. STYROFOAM CUP
$\frac{19\pi}{3}$ or 19.9 in.^{3}

14. RED ROAD
82.5 square units

CHAPTER 7, CLASSIC PROBLEMS

16. NINE COINS
Balance 3 vs. 3. If it balances, then use the other 3 and weigh 1 vs. 1. If the first weighing does not balance, then take the lighter 3 and weigh 1 vs. 1. When 1 vs. 1 balances, it's the third coin. When 1 vs. 1 does not balance, then it's the lighter one.

17. TWELVE COINS
Start by balancing two groups of 4.

Case 1. If they balance, then the counterfeit is in the remaining 4. Balance 3 of the remaining 4 against 3 of the coins now known to be okay. If they balance, then the fourth remaining coin is counterfeit. Balance it against a known weight, and you will know if it is heavier or lighter. If 3 unknown vs. 3 known didn't balance, then you will know whether the counterfeit is heavy or light. Balance 2 of the 3 suspect coins against each other. If they balance, the third suspect is the counterfeit. If they don't, then the heavier or lighter, based on weighing #2, is the counterfeit.

Case 2. If the original two groups of 4 didn't balance. Identify these three groups of 4 as possibly heavy—H_1, H_2, H_3, H_4; possibly light—L_1, L_2, L_3, L_4; and okay—O_1, O_2, O_3, O_4. Now balance 4 "okays" and 1 "heavy" against 3 "heavies" and 2 "lights."

$$O_1 \, O_2 \, O_3 \, O_4 \, H_1 \text{ vs. } H_2 \, H_3 \, H_4 \, L_1 \, L_2$$

There are three possible outcomes:

Case 1. They balance: For weighing #3, test L_3 and L_4 against each other. Whichever is lighter is the counterfeit coin, and it is light.

Case 2. Heavy on the left. So, either H_1 is heavy *or* L_1 or L_2 is light. Test L_1 against L_2. If they balance, then H1 is heavy. If they don't balance, then whichever is lighter is the counterfeit.

Case 3. Heavy on the right. So, H_2, or H_3, or H_4 is heavy. Test H_2 vs. H_3. If they balance, it's H_4. If not, then it's the heavier.

Chapter 7, Problem Set A, Version 2

1. MAGNET SCHOOL
$16\,^2/_3\%$

2. ORANGES
10 oranges

3. BOYSENBERRIES
$5.76

4. STEREO SALE
$127.58 (or $127.575)

5. LIFE'S NECESSITIES
8 sodas

6. AN A IN MATH
She can't. (She would have to get 104.2%.)

7. SOIL AND SAND
40 cubic yards

8. ONE HUNDRED SIXTY-EIGHT INCHES OF STRING
360 square inches

9. JUAN'S CHURCH
10.9375%

10. A LOT OF STUFF
7 pounds 13 ounces

11. SHOOTING PERCENTAGE
60% (12 out of 20)

12. **CAMP STOVE**
22 minutes

13. **ROAD RALLY**
32.7 miles per hour

14. **CLASSIC BOOKS**
48 books on third shelf

15. **LONG ROAD**
48 miles long

16. **CAT LOVES BEANIE BABIES**
333 Beanie Babies in the family room

17. **HONOR SOCIETY**
70 students from Roseville

Chapter 7, Problem Set B

1. **WHO WEIGHS WHAT?**
Morris, 180; Frank, 168; Pedro, 173;
Devon, 155; Fua, 164

2. **FAMILY DAY**
23 sections (or 22 sections and 4 rows)

3. **CARROT JUICE**
112 quarts

4. **HRUNKLA APARTMENT HOUSES**
   a. Ride belt to SE corner.
   b. Ride to eighth floor (or fourth or twelfth).
   c. Ride north to NE corner.
   d. Ride to seventh floor (or third or eleventh).
   e. Ride W to NW corner.
   f. Ride to ninth floor.
   g. Ride E to friend's door.
   Note: Different routes are possible.

5. **NIGHTMARES**
October 31 (Boo!)

Chapter 7, Problem Set B, Version 2

1. **USED CARS**
Lemons cheap and no green people

2. **THE NATURAL LOOK**
$11/26$

3. **TYPESETTING**
He runs out after 162.

4. **A FAMILY MAN**
Daughter, 23; son, 18; man, 45

5. **HOW TO AVOID FALLING ASLEEP AT A MEETING**
There were 25 people; 1 left after counting by 6's; 1 left after counting by 7's; 1 came back after counting by 8's; 1 came back after counting by 10's.

Chapter 7, Problem Set B, Version 3

1. **WHO'S ON THE BENCH?**
There were 17 people on the bench. The team came in after counting by 3's.

2. **GAMUSE PLAYGROUND**
Take slides down to ninth, then seventh, then fifth floor. Take tube up to eighth floor. Take escalators up to tenth, then twelfth floor.

3. **BICYCLE TRAINING**
20 miles per hour

4. **SIERRA SLUGGERS**
K. C., 150; Mark, 138; Larry, 149; Jack, 131; Chris, 140

5. **PRIZE MONEY**
They win $25,963 each.

Chapter 7, Problem Set B, Version 4

1. **MILEAGE SIGN**
   61 miles

2. **A NEW VERSION OF SCRABBLE**
   A = 2, E = 1, I = 4, O = 6, U = 3,
   B = 11, G = 1, L = 2, M = 4, N = 5,
   P = 6, R = 10, S = 3, T = 9, V = 14,
   W = 8; PROBLEM-SOLVING = 986

3. **WIDE SLIDE**
   504 seconds later

4. **THREE-DIMENSIONAL CHESS**
   It takes three moves. (There are many
   ways to make the moves.)

5. **LEAKY SINK**
   21.5 seconds

Chapter 8, Problem Set A

1. **SODA CALORIES**
   About 100 pounds/year

2. **CHRISTINA'S TRIP**
   a. 51.6 mi/hr   f. 0.12 qt/min
   b. 29 mi/gal   g. 4.8 cents/min
   c. $1.63/gal   h. 5.6 cents/mi
   d. 75.6 ft/sec   i. 17.8 mi/dollar
   e. $2.90/hr   j. 1.78 gal/hr

3. **UNIT CONVERSIONS**
   a. 114.8 ft   f. 34.3 in.
   b. 51.8 m   g. 1371.5 mm
   c. 241.4 km   h. 14.3 m/sec
   d. 29.2 mi   i. 18.95 L
   e. 121.9 cm   j. 16.9 qt

4. **ANOTHER LONG COMMUTE**
   a. 2.5 gal   e. 4.57 cents/mi
   b. 1.75 hr   f. 262.5 pass-mi
   c. $1.33/passenger   g. 105 pass-mi/gal
   d. 73.3 ft/sec   h. 1.52 cents/pass-mi
   Gerónimo's car is more efficient:
   1.14 cents/pass-mi vs. 1.52 cents/pass-mi.

5. **IN-HOME SODA MACHINES**
   11 cents per liter, or 23 cents per six-pack

6. **PAINTING CHIPMUNKS**
   Alvin, $24; Simon, $19; Theodore, $18

7. **FIREPLACE INSERT**
   She will spend $138.60 instead of $500, so
   will save $361.40.

8. **NEW CAR PURCHASE**
   $206.25

9. **LEAKY FAUCET, PART 2**
   2190 gal/year

10. **SPEEDING UP**
    56.25 miles per hour

11. **READING RATE, PART 1**
    0.44 page per minute

12. **READING RATE, PART 2**
    About 20 or 21 books (20.5)

13. **FLOODS**
    507,621.51 seconds (5 days 21 hours)

14. **PENALTY KICK**
    0.3 second

15. **SAILING SHIPS**
    50.6 feet

16. **NURSING**
    8.4 cubic centimeters per hour (would
    probably need to be a whole number,
    either 8 or 9)

17. **DRIVE OR FLY: WHICH IS SAFER?**
    0.716532 death/billion passenger-miles for
    flying

    11.5882 deaths/billion passenger-miles for
    driving

    If driving is as safe as flying, then 2601
    deaths/billion passenger-miles (instead of
    42,065)

19. THE TELEGRAPH POLES

319 ft/pole

Chapter 8, Problem Set A, Version 2

1. MORE CONVERSIONS

a. 147.6 ft   f. 9.4 in.

b. 85.3 m   g. 1219.1 mm

c. 402.3 km   h. 22.4 m/sec

d. 19.3 mi   i. 11.4 L

e. 1066.7 cm   j. 21.1 qt

2. LIQUID PIPELINE

Answers will vary. Here is one possible answer:

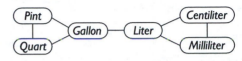

3. TIME IN A PIPELINE

Answers will vary. Here is one possible answer:

4. WHAT'S UP, DOC?

a. $0.07/carrot   c. 3.2 ounces/carrot

b. 7 cents/carrot   d. 5 carrots/pound

5. MASON'S TAXI SERVICE

a. 33 gal   f. $15.26/pass

b. 150 mi/hr   g. 11.1 cents/mi

c. 220 ft/sec   h. 2200 pass-mi

d. $16.65/hr   i. 66.7 pass-mi/gal

e. 16.7 mi/gal   j. 2.8 cents/pass-mi

6. MARCEL'S TRIP

a. 14 gal   f. $2.71/pass

b. 7 hr   g. 3.9 cents/mi

c. 88 ft/sec   h. 2520 pass-mi

d. $2.32/hr   i. 180 pass-mi/gal

e. $1.16/gal   j. 0.64 cent/pass-mi

7. COMMUTER FLIGHT

a. 9.2 mi/min   d. 44,000 pass-mi/hr

b. 806.7 ft/sec   e. 733.3 pass-mi/min

c. 32,000 pass-mi

8. A SHOT IN THE DARK

a. 134.2 mi/hr   b. 216 km/hr

9. IT'S ABOUT TIME

Each person gets $30 except Jeannette, who gets $36.

10. AFTER THE DEADHEADS

Sassafras, $27.37; Moonshine, $14.49; Peace, $19.32; Harmony, $28.98; Chynna, $38.64

11a. WHERE'S THE RUE?

6 miles per hour

11b. WHERE'S THE VACHE?

95.8 times per year

12. MANUSCRIPT

357 words per minute; 43 pages per hour

13. EVENING EXERCISE

20 holes per walk

14. AXELROD'S TRIP

a. 52.8 mi/hr   d. 77.4 ft/sec

b. $18.72   e. 4.73 cents/mi

c. 27.5 mi/gal

15. ZEKE AND FRIENDS

a. 244 mi
b. $1.23/gal
c. 48/8 mi/hr
d. 122 pass-mi/gal
e. 1.01 cents/pass-mi

16. HORSE RACING
67.5 seconds

Chapter 8, Problem Set B

1. SPICE ON ICE
17 adults, 13 juniors, 90 children

2. VOLLEYBALL LEAGUE
Buckeyes beat Bombay Bicycle and lost to Sacto Magazine and Red Skeletons.

3. QUIT WHILE YOU CAN
14 years 213 days 1 hour and 30 minutes (using 365.25 days/yr)

4. NO CHAIN LETTERS
98,415 letters

5. WHO WAS SNOOZING?
Charlton was the snoozer.

Chapter 8, Problem Set B, Version 2

1. I CAN SEE FOR MILES AND MILES
1 hour 41 minutes 15 seconds

2. AN AGE-OLD PROBLEM
Kevin is 35, Kate is 51, and Lee is 57.

3. CROSS-NUMBER PUZZLE 1

3	8	1
7	8	4
7	2	9

or

2	6	4
3	2	4
3	4	3

4. FILLING THE PLANTER BOX
$26^{2}/_{3}$ trips, or 27 complete trips

5. GENEROUS FRIENDS
Each glass was half full.

Chapter 8, Problem Set B, Version 3

1. DISCOUNT HARDWARE
3 hammers, 11 screwdrivers, 86 nails

2. LIFE SPAN
73 years 343 days

3. COIN COLLECTION
2 dimes and a nickel

4. THE GOAT PROBLEM
2156.25π square feet, or approximately 6774 square feet

5. CROSS-SUMS

Chapter 8, Problem Set B, Version 4

1. WALLY WEIRD'S WINE
For the twentieth century: 1908, 1917, 1926, 1935, 1944, 1953, 1962
For the twenty-first century: 2007, 2016, 2025, 2034, 2043, 2052, 2061, 2079

2. **LAKE MEAD FLOODS THE MOJAVE DESERT**
3.26 feet

3. **CROSS-NUMBER PUZZLE 2**

6	1
4	7

4. **VCR**
Part 1: Record at LP; will have 10 minutes left.
Part 2: Switch after 10 minutes.

5. **SPRINKLERS**
57%

Chapter 9, Problem Set A

1. **DIAGONALS**
275 diagonals

2. **SUM OF ODDS**
25 million

3. **TV TRUCK**
504 TV sets when all sets are arranged facing the same direction according to height; 528 TV sets when all sets are arranged at the same height but with some sets sitting with length forward and others sitting with width forward

4. **POTATOES**
63 pounds of potatoes; 7 soldiers

5. **SQUARE AND HEXAGON**
$$h + p = \frac{T}{6} + 4S$$

6. **ODD AND EVEN**
500

7. **CHINESE NEW YEAR**
The Year of the Monkey

8. **TWENTY-FIVE-MAN ROSTER**
12 infielders/outfielders, 6 starting pitchers, 4 relievers, 3 catchers

9. **LAST DIGIT**
6

10. **REMAINDER**
The remainder is 3. It cycles through 12 remainders.

11. **FIFTY-TWO-CARD PICKUP**
$2^{52} - 1$

Chapter 9, Classic Problems

14. **PAINTING THE LAMPPOSTS**
Pat painted six more.

15. **BILLIONS**
49,499,999,995,500,000,000

Chapter 9, Problem Set A, Version 2

1. **LAST DIGIT AGAIN**
The last digit is 1.

2. **DIAGONALS OF A POLYGON**
77 diagonals

3. **SUM NUMBERS—LOTSA NUMBERS**
96,012,000

4. **AIR FARE**
1 hour 50 minutes

5. **SEASON TICKET PLANS**
$2^{42} - 1$

6. **ONCE A YANKEES FAN, ALWAYS A YANKEES FAN**
$2^{25} - 1$

7. **CLASSIC MUSTANGS**
Travis, 6; Sandra, 3; Narrator, 2

8. VIVE LA DIFFERENCE!
125,750

9. SAVING PENNIES
Maritza saved $19.84 more than Ricardo.

10. BUSINESS CARDS
132,860

Chapter 9, Problem Set B

1. COVERING THE GRID
30 tiles

2. NINE POINTS
84 triangles

3. BASKET PARTIES
20,475 parties

4. ADDING CHLORINE
1 1/4 ounces

5. JOGGING AROUND A TRACK
Every 40 seconds

Chapter 9, Problem Set B, Version 2

1. COVERING THE PATIO
401 nails

2. LIKE A WILDFIRE
16,376 people (including the 8 firefighters who started the program)

3. SUM OF TEN
282

4. LAPPING JOGGERS
Jan ran 3 miles, and Silvia ran 3.75 miles.

5. THE PHOON BROTHERS
Each Phoon pays $11.73 and Helen pays $3.91, or Phoons each pay $11.95 and Helen pays $3.25. Some rounding is needed to ensure that all the money gets paid. Answers can be pennies different, as long as total amount paid is $39.10.

Chapter 9, Problem Set B, Version 3

1. PYRAMID SCHEME
24,570 investors

2. LOSING TIME
20 days later on a Saturday

3. DRIVE ME WILDE
Fawn, Kat, and Wolfgang each pay $5.19; Duckie pays $4.02; Mildred pays $0.75. Other answers: Fawn, Kat, and Wolfgang each pay $5.07; Duckie pays $4.19; Mildred pays $0.94. Some rounding is necessary to ensure that all the money gets paid. Answers can be pennies different as long as total paid is $20.34.

4. SOCCER LEAGUE

Team	Wins	Losses	Ties
A	1	8	1
B	5	5	0
C	10	0	0
D	5	4	1
H	7	2	1
T	0	9	1

Allosaurus tied Triceratops. Dimetrodon tied Hypsilophodon. Brontosaurus and Dimetrodon split their games.

5. RUNNING ERRANDS
120 different sets of errands

Chapter 9, Problem Set B, Version 4

1. TWO VEHICLES
The van has driven 17,000 miles.

2. DIGITAL CLOCK
5:02

3. WATCH YOUR SPEED
60 miles per hour

4. TWO-LANE HIGHWAY
100 passing periods

5. PEDESTRIAN
28 miles per hour

Chapter 10, Problem Set A-1

1. THREE ADULTS AND TWO KIDS
13 trips, starting with the two kids going across

2. THE DOG, THE GOOSE, AND THE CORN
There are two solutions, which are virtually identical:

He must take the goose across first, leave it, and go back for the corn. He then takes the goose back to the first side, leaving it and taking the dog across. He leaves the dog with the corn and returns across the river to retrieve the goose.

Or, he must take the goose across first, leave it, and go back for the dog. He then takes the goose back to the first side, leaving it and taking the corn across. He leaves the corn with the dog and returns across the river to retrieve the goose.

3. HOOP GREETING
45 handshakes

4. SWITCHING JACKALS AND COYOTES
5 trips; the boat starts with the jackals.

5. THE HOTEL BILL
The bellhop had $2 that the women had paid the hotel. The hotel had $55 from the women. The women each paid $19, and out of that total of $57, the bellhop has $2 and the hotel has $55. There is no missing dollar.

6. PERSIS'S GIFT SHOP
Persis lost either $25 or $32, depending on whether you count the cost of the figurine at $6 or $13.

Chapter 10, Problem Set A-2

7. TWO JACKALS LOSE THEIR LICENSES
13 river crossings. The key to this problem is trading the jackal who can row for one of the other jackals in the middle of the problem.

8. JACK-QUEEN-DIAMOND
Jack of diamonds, queen of hearts, and queen of diamonds

9. BASEBALL SEATING
Aisle, Mom, Alyse, Dad, Kevin, Jeremy (or put Mom on the other side of the aisle)

10. MAGIC TRIANGLE

11. MAGIC SQUARE

There are a number of possible solutions. One possible solution is shown below. Some of the others can be found by switching columns with columns or rows with rows.

1	8	3
6	4	2
5	0	7

12. TRUE EQUATIONS

Switching the 4 and 8 with the 3 and 6, respectively, yields another solution:

$$2 + \boxed{7} = \boxed{9}$$

$$\boxed{5} \times \boxed{6} = \boxed{3}\,\boxed{0}$$

$$\boxed{8} \div \boxed{4} = 2 \qquad \boxed{2} - 1 = \boxed{1}$$

13. THREE-ON-THREE BASKETBALL

	Team 1	Team 2	Team 3
Center	Horace	Ingrid	Jerome
Forward	Sasha	Kedra	Taunia
Guard	Leon	Weston	Kathryn

Taunia is the forward on Kathryn's team. Leon plays guard on Sasha's team. Kedra plays forward for a team including Ingrid and Weston.

14. CUBIST

E, right side up in normal orientation (*Note*: I appears twice.)

15. FOLDING CUBES

Answers will vary.

16. CO-ED VOLLEYBALL

The problem as stated contains contradictory information. One or more constraints will have to be ignored.

17. WORCHESTERSHIRE PALACE

a. 16 moves   b. 14 moves

18. PROBLEM VATS

There's the same amount of water in the alcohol as there is alcohol in the water.

CHAPTER 10, CLASSIC PROBLEMS

20. CROSSING A BRIDGE WITH A FLASHLIGHT

They got across in this order:
1. Women 1 and 2 go across, Woman 1 returns.
2. Women 3 and 4 go across, Woman 2 returns.
3. Women 1 and 2 go across.

21. STEALING THE CASTLE TREASURE

Assumption: All are up in the tower.
They escaped in this order:
1. Treasure down
2. Treasure up, boy down
3. Boy up, youth down
4. Treasure down
5. Youth and treasure up, man down
6. Treasure down
7. Treasure up, boy down
8. Boy up, youth down
9. Treasure down
10. Treasure up, boy down
11. Treasure down

Chapter 10, Problem Set A, Version 2

1. HANUM'S ISLAND

9 trips

2. ANOTHER MAGIC TRIANGLE

There are several possibilities. One is shown below.

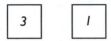

3. ROOKIE PURCHASE

Max lost $1,500.

4. LAWN CHAIR AND BARBECUE

Marc made $2, Wilma lost $1, and presumably Sammie lost $1, though he currently owns the chair and grill and could make or lose more money on them.

5. STAMPS

7 different ways

6. LETTER CUBE

Place the letters as shown on the folded-out cube. The answer is N (oriented right side up).

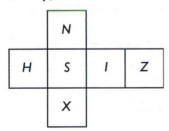

7. SIX AND FOUR

3 rotations on the left die, 3 rotations on the right die

8. A HEART IN THE RIGHT PLACE

Two of diamonds, three of hearts, and six of diamonds

9. TWO DICE

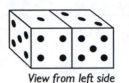 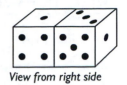

View from left side   *View from right side*

10. JUST A PEW KIDS

This is impossible, unless you want to split the family across an aisle.

11. FIVE AT THE TABLE

There are 20 ways, or 45 ways if you consider that position within the table matters. (Reflections are considered different arrangements.)

12. ANOTHER MAGIC SQUARE

2	6	7
10	5	0
3	4	8

13. CARDS IN TWO ROWS

From left to right, there is a jack with a queen on top of it, an ace with a king on top of it, and a queen with an ace on top of it.

14. BLUEGRASS AT DINNER

Fiddle	Banjo	Guitar
Emily	Ben	Lucille
Sue	Marty	Kathleen
George	Irving	Kris

Sue plays fiddle in Marty's group. Ben plays banjo in Emily's group. George plays fiddle, and Irving and Kris are also in his group.

15. GOOD NEWS FOR CUBES

All three combinations form cubes.

Chapter 10, Problem Set B

1. **DECREASING NUMBERS**
1013

2. **WHITE SALE**
She bought 18 washcloths and 12 towels. All of these items would have cost $65.40 if she had bought them on Saturday. She shouldn't have been so happy.

3. **KMRCL TV**
43.4 hours per week, or
43 hours 24 minutes

4. **THUNDER AND LIGHTNING**
In 5 seconds, sound travels 1.028 miles. The rule of thumb is within 3% accuracy.

5. **STATE FAIR**
From a visitor's perspective—this is right to left:
Foot Massage, empty (ladder sellers), Hot Spas, Computer Horoscope, Encyclopedia Antarctica, empty (vacuum sellers), Slice-It-Dice-It Veggie Peeler
So Dad is in the third booth from the right or the fifth booth from the left.

Chapter 10, Problem Set B, Version 2

1. **DUELING PINS**
Lani and Hank Wojic, 290 and 217 pins; Fawnda and Bill Carson, 258 and 211 pins; Rhonda and Derrek Sanchez, 281 and 243 pins. Wojics beat Sanchezes 290 to 243; Sanchezes beat Carsons 281 to 211; Carsons beat Wojics 258 to 217.

2. **STOPLIGHTS**
You will have to stop at the 18th light after you started (counting the starting light as number 1), so you can make it through 17 intersections (including the first intersection).

3. **TICKET PRICES**
$7, $10, $16, $28

4. **CAPICUA NUMBERS**
1098 or 1089 (not counting 1 to 9)

5. **A FAST (?) SPACESHIP**
308 years

Chapter 10, Problem Set B, Version 3

1. **ODD-ODD NUMBERS**
19,530

2. **THE BRIDGE TOURNAMENT**
Dave and Mary Hatfield, 880 points; Jim and Jean Friedrich, 680 points; Anne and Randy Muir, 490 points

3. **DISCOUNT TICKETS**
22 children's tickets

4. **DRIVING ON THE FREEWAY**
8.8 car lengths

5. **CAKE RECIPE**
The volume ratio is 1.23. To make the eggs work, use 1.25. So, all amounts are $1\frac{1}{4}$ times the original amounts. The new amounts are as follows:
$3\frac{1}{3}$ cups cake flour
$2\frac{13}{16}$ teaspoons baking powder ($2\frac{3}{4}$ is more realistic)
$1\frac{1}{4}$ cups butter
$2\frac{1}{2}$ cups sugar
5 eggs
$1\frac{1}{4}$ cups milk

Chapter 10, Problem Set B, Version 4

1. **BATTING AVERAGE**
.152 during vacation

2. **PYRAMIDS**
Listed from lightest to heaviest: Lori, Kelly, Diane, Debbie, Barbara, Carmella, Lisa, Kate

3. **STEP ON A CRACK**
$3/11$

4. **START YOUR ENGINES**
Perfect squares 1, 4, 9, and so on, to 144

5. **RODGERS AND HAMMERSTEIN**
Press "Disc Skip" once, then once, then twice, then twice, then twice, then twice, exchanging each time.

Chapter 11, Problem Set A

1. **LOSING STREAK**
$9,200

2. **GENEROSITY**
$252

3. **COOKIES**
He started with 127 cookies. He ate 115, which translates into 11.5 pounds.

4. **THE MALL**
She started with $124. She spent $18 on a CD, $53 on a dress, $11 for lunch, $14 on a book, $12 for gas, and $4 on a tape. She gave her sibling $2.

5. **USED CAR**
$6,000

6. **JASMINE'S DANCE TROUPE**
29 members

7. **WHAT'S MY NUMBER?**
6

8. **LOST HIS MARBLES**
20 marbles

9. **HOCKEY CARDS**
Jack 25, Jill 15

10. **TWO FOR TENNIS**
34 balls

11. **DONUTS**
19 donuts

12. **GOLF CLUBS**
13 golf clubs

Chapter 11, Classic Problems

14. **THE THREE BEGGARS**
42¢

15. **X AND O**
O wins

Chapter 11, Problem Set A, Version 2

1. **MARY'S CAR LOT**
20 cars

2. **ORANGE YOU HUNGRY?**
36 oranges

3. **TURNPIKE**
$13

4. **THE MONOPOLY GAME**
$250

5. **BROWN THUMB**
23 plants

6. **JELLY BEAN RABBITS**
30 jelly beans: Lily ate 17; I ate 13.

7. **WINNING GOLDFISH**
5 goldfish

8. **PRETTY FISHY**
26 fishbowls

9. **POLITICAL BUTTONS**
65 buttons

10. **FRUIT STAND**
39 watermelons

Chapter 11, Problem Set B

1. **HOW MUCH DOG FOOD?**
   $74.40

2. **THE LUGGAGE RACK**
   $2.75 more, and it would take 18 extra minutes.

3. **CROSSING THE RIVER WITH DOGS**
   13 trips. Odd-numbered trips cross over; even-numbered trips cross back.
   1. L-dog, M-dog, P-dog
   2. L-dog
   3. L-dog, J-dog
   4. L-dog
   5. Mama, Papa, Judy
   6. Mama, M-dog
   7. Lisa, L-dog
   8. Papa, P-dog
   9. Papa, Mama, Ed
   10. L-dog
   11. L-dog, E-dog, M-dog
   12. L-dog
   13. L-dog, P-dog

4. **DON'T FEED THE ANIMALS**
   62 bags of peanuts

5. **LOST IN PURSUIT OF PEANUTS**
   North, 45 hours, 10 full bags (16 peanuts left over)

Chapter 11, Problem Set B, Version 2

1. **CHICKEN EGGS**
   48 eggs

2. **HEARTBEATS**
   4,828,950 more heartbeats for Ken in a year

3. **PENNIES IN THE SAND**
   Started with 50 pennies, lost 53

4. **JOGGING**
   Between 5.0 and 5.2 miles per hour

5. **THE BRIDGE OVER RAT-TONGUE RAVINE**
   7 trips. Odd-numbered trips cross over; even-numbered trips cross back.
   1. Thunder and Eagle
   2. Eagle
   3. Three Rattlers and Dragons
   4. Thunder
   5. Three Rattlers and Eagle
   6. Eagle
   7. Thunder and Eagle

Chapter 11, Problem Set B, Version 3

1. **HOT DOG**
   48 hot dogs

2. **FRATERNITY OUTING**
   11 trips (there is more than one possible way to arrange the 11 trips). Odd-numbered trips are trips to the contest; even-numbered trips are trips back.
   1. J, TR, BZT, NK
   2. J, TR, NK
   3. J, TR, EP
   4. J, TR
   5. J, TR, 2PPP
   6. J, TR, BZT
   7. J, TR, BZT, 1PPP
   8. J, TR, BZT
   9. J, TR, 2PPP
   10. J, TR
   11. J, TR, BZT, NK

3. **WHICH CAR SHOULD THEY TAKE?**
   $4.95 one way; $9.90 round trip

4. **HOT DOGS AGAIN**
   45 hot dogs

5. **LOST IN GRIDLOCK**
   2 blocks east, 5 blocks south

Chapter 11, Problem Set B, Version 4

1. **LOST CARDS**
   140 cards

2. **FENCE PAINTING**
$23\frac{1}{3}$ fences

3. **MARK AND DIANE**
330 different routes

4. **WALKING**
287.46 feet

5. **WEDDING RECEPTION**
There are many possible ways. Here's one:
1. J. R., Mina, Brian, Jerri
2. Jerri
3. Len, Phil, Jerri
4. Jerri
5. Amanda, Kevin, Walter, Jerri
6. Brian, Jerri
7. Jerri, Brian, Jill, Lincoln

Chapter 12, Problem Set A

1. **MORE CATEGORIES**
   1. Trumpets, pianos, musical instruments, clarinets, violins, trombones, brass, woodwinds

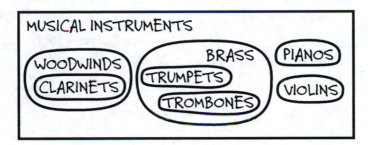

2. Water vessels, submarines, war boats, sailboats, battleships, ferries

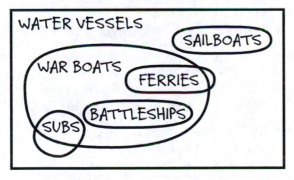

3. Old cows, dairy cows, cows, cows wearing bells

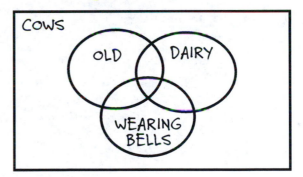

2. **NORTHERN ATLANTIC MUSIC COLLEGE**
45 students

3. **HAMBURGERS AND HOT DOGS**
45 liked neither.

4. **ROCK BAND**
11 band members

5. **TEACHER CANDIDATES**
22 students

6. **EATING VEGETABLES**
1, 4, 4, 0, 6 children

7. **THE FIELD TRIPS**
88 children

8. **NO PETS ALLOWED**
2 residents

9. **SPORTSMEN**
51 men

10. FAMILY REUNION
5 cousins were neither nieces nor aunts.

11. JUST WHAT ARE THESE THINGS, ANYWAY?
53 ENAJs are neither DERFs nor SIVADs.

12. FLU AND YOU
a. 148 people
b. 26 people
c. 46/98 = 47%

13. BLOOD LINES
76, 12, 4, 24 patients

14. CAMPUS LIFE
a. 9 students
b. 4 students

Chapter 12, Problem Set A, Version 2

1. FORMAL DANCE
36 students

2. SODA SURVEY
209 people

3. COMPANY BARBECUE
a. 92 people
b. 35 people
c. 27 people
d. 5 people
e. 20 people

4. SPIRIT OF WOODSTOCK
1 member

5. COUNTRY CLUB
115 members

6. EAST PARKING LOT
18 white vehicles

7. SPLIT TICKET
70 people

8. BRAVE ALL-STARS
36 people

9. EARTHQUAKE DAMAGES
4 houses; 4 houses

10. MARKETING COMPANY DOES SODA SURVEY
a. 90 people
b. 31 people
c. 21 people
d. 20 people
e. 13 people

11. HEALTH CLUB ACTIVITIES
a. 63, b. 16, c. 10, d. 32, e. 9

Chapter 12, Problem Set B

1. HOLIDAY PASTRIES
84 total: Papa 30, Ed 15, Lisa 15, Judy 10, Mama 4, dogs 10

2. PRESENTS! OH BOY!
Judy, 59", plastic bag, Legos; Gail, 62", pillowcase, shoes; Keith, 65", newspaper, picture; Lisa, 68", box, belt; Ed, 71", towel, candy

3. LEGO MY PYRAMID
Bumps showing, 400; Legos used, 385

4. THE HOLIDAY PARTY
30 cousins were neither nephews nor uncles; 10 uncles were cousins but not nephews.

5. HOLIDAY DINNER
19 meals

Chapter 12, Problem Set B, Version 2

1. THE TREASURE OF SIERRA MARBLES
Tom, 69; Huck, 51; Becky, 40

2. TOOTHPICK SQUARES
220 toothpicks

3. THE FRESHMAN CLASS
14 students

4. **THE SANDWICH SHOP**
152 different types of sandwiches

5. **FINE ARTS**
Alan Innis, symphony, publisher;
Chris Hatfield, musicals, reporter;
Bev Gunderson, jazz, engineer;
Doreen Jackson, art, doctor;
Ernie Fillmore, ballet, lawyer

Chapter 12, Problem Set B, Version 3

1. **POLITICAL PARTIES**
Maria Lamson, bus, the Bronx;
Walter Peterson, bike, Queens;
Valerie Archer, car, Seaside;
Jasper Doyle, walk, Manhattan;
Teresa Sinderson, subway, Long Island

2. **COUNTRY MUSIC**
8 songs

3. **HUNGRY BROTHERS**
55 cookies

4. **LET GO MY LEGO**
Legos, 220; bumps visible, 440

5. **HOW MUCH STEREO CAN YOU AFFORD?**
18 different ways

Chapter 12, Problem Set B, Version 4

1. **FOUR FRIENDS**
Driver, Areatha, soup, lemon-lime; front
seat passenger, Marco, spaghetti, cola;
back seat left, Sara, chili, orange soda;
back seat right, Roberto, macaroni and
cheese, root beer

2. **SOCCER TEAM**
Four people play goalie. Three people
play fullback and forward only.

3. **ORANGE TREASURE**
160 oranges at start; Ahab, 73;
Bluebeard, 46; Hook, 35; monkeys, 6

4. **PENTAGON**
35 triangles

5. **DIGIT 8**
600,000

Chapter 13, Problem Set A

1. **ALGEBRA THIS TIME**
Answers will vary.

2. **MORE COINS**
27 nickels, 19 dimes

3. **SUPPLEMENTS**
58° and 122°

4. **BIKE RIDE**
$13\frac{1}{3}$ miles per hour

5. **CHAMPIONSHIP GAME**
179 students

6. **FISHING POLES**
Daniel, 8 poles; Gary, 5 poles

7. **CAR WASH**
12 minutes

8. **INTEREST**
$7,567.57 (Any answer between
$7,567.30 and $7,567.83 rounds to the
correct amount of interest.)

9. **CHEMISTRY**
29.41 gallons of 24%, and 20.59 gallons
of 41%

10. **TICKET PRICE INCREASE**
7566 with card; 4872 without card

11. **LADDER**
About 16 feet

12. **MADAME XANADU'S CRYSTAL BALL**
The ball will be safe because the distance from the edge of the table to the corner is approximately 9.9 inches, and the diameter of the ball is 10 inches.

13. **STUPID NUMBER TRICK**
This trick works in 2003. Algebra reveals the result. $100x + 103 - y$ or $100x + 102 - y$, where x is the number of baths and y is the age in 2003.

14. **TREADMILL**
   a. 10 minutes 40 seconds walking, 4 minutes 20 seconds jogging
   b. 114 beats/minute

15. **THOSE DARN BILLS KEEP PILING UP**
Utilities: $43; rent: $215; phone: $68; food: $170; credit card: $255

CHAPTER 13, CLASSIC PROBLEMS

17. **BLENDING THE TEAS**
70 pounds of 32¢ tea and 30 pounds of 40¢ tea

18. **A DIESEL SHIP AND A SEAPLANE**
200 miles from shore. Note that the actual time and speed cannot be determined.

19. **INVERNESS TO GLASGOW**
$82\frac{11}{16}$ or 82.6875 miles from Glasgow

Chapter 13, Problem Set A, Version 2

1. **MORE AND MORE COINS**
11 quarters, 7 dimes

2. **COMPLEMENTS**
29° and 61°

3. **MARYLOU'S INVESTMENTS**
$5,000 at 8%, and $4,000 at 5%

4. **A SMALL WOODWORKING COMPANY**
17 small packages and 9 large packages

5. **SPEEDING TIX**
The speed limit was 25 miles per hour. Chester was going 48 miles per hour.

6. **DAD GETS LOTS OF HELP**
$93\frac{1}{3}$ minutes

7. **MOM GETS HELP?**
−120 minutes (What does a negative answer mean in a problem like this?)

8. **TAXING BERNICE**
$40,500 was taxed at 7%, and $1,500 was taxed at 9%.

9. **TERRY'S BOO-BOO**
$1\frac{3}{7}$ gallons of concentrate

10. **ALGEBRA AREA**
595 square centimeters

11. **MICHAELA'S SEASON**
17 singles, 3 doubles, and 4 triples

12. **GOLD COUNTRY TIRE**
29 customers bought 2 tires. 6 customers bought 4 tires.

13. **CLARENCE AND STEPHANIE ARE THINKING**
Clarence, 16; Stephanie, 6

14. **SUGARY PUNCH**
Add 4.4 liters of water.

15. **SLED RUN**
About 293.2 feet

16. **FEDERAL STANDARDS**
50,000 cars

17. **ANOTHER COIN PROBLEM**
37 dimes, 23 nickels

18. **STILL MORE COINS**
27 dimes, 57 quarters

19. DECENT INVESTMENTS
$5,018.20

20. INVESTMENTS
$4,254.90 in the 1.7% account

21. VISITING A FRIEND
36 miles

22. ROAD TRIP
1.45 hr

23. RACHAEL'S JUICE
23 quarts

24. TALIA'S TRULY TERRIFIC FRUIT PUNCH
7.36 quarts

Chapter 13, Problem Set B

1. HOW MANY ZEROS?
1249 zeros

2. FORMING PENTOMINOES
12 different pentominoes

3. VALLEY SPRINGS
4 to 11

4. ALL IN THE FAMILY

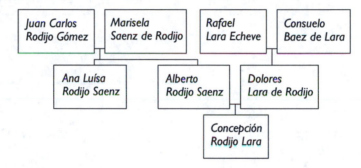

5. WHEN I'M 64
Start with 1, then make sure any subsequent choice of yours and of your opponent add up to 9.

Chapter 13, Problem Set B, Version 2

1. THE DRAMA PRODUCTION
(A = Augie, B = Beatrice, and so on)
A, G, B, J, C, H, D, L, E, I, F, K

2. SELLING STEAK
Karl is charging $18.75, and there are 25 boxes.

3. DICEY PRIMES
15 to 13

4. ICE CREAM COUPONS
Both were juniors.

5. HALFWAY LINES
Switch the 7 and the 10, draw a vertical line between the second and third columns; the horizontal line goes below the second row.

Chapter 13, Problem Set B, Version 3

1. LATE TO THE GATE
4.5 miles per hour

2. CARD ARRANGEMENT
From top: 1, 6, 2, 10, 3, 7, 4, 9, 5, 8

3. RECTANGLE RATIOS
Length: 15; width: 3

4. **TRIOMINOES**

12 different figures

5. **NEMATOAD**

Match every pickup of beans with your opponent's to total 5. (For example, if the opponent picks up 3, you pick up 2.)

Chapter 13, Problem Set B, Version 4

1. **AMAZING NUMBERS**

Right to 4, diagonally down right to 6, left to 1, diagonally up right to 4, down to 5, right to E

2. **LID ASTRAY**

There are several different strategies. Look for efficiency. Here is an efficient way requiring no more than three trials:

Label containers small to large C_1, C_2, C_3, and lids small to large L_1, L_2, L_3, L_4, L_5. Try L_3 on C_2.

If L_3 is too big for C_2, then L_1 fits C_1 and L_2 fits C_2. If L_3 is too small for C_2, then L_4 fits C_2 and L_5 fits C_3. In either case you only need one more trial and you will be done.

If L_3 fits C_2, then you need two more trials.

3. **RADIATOR**

4.16 quarts

4. **LSAT TEST**

14 arrangements

5. **LARGE POWER OF TWO**

797

Chapter 14, Problem Set A

1. **EIGHT FUNCTIONS**

a. $y = 7x + 6$
b. $y = x^2 + 5x + 6$
c. $y = 2x^2 - 3x + 4$
d. $y = -4x + 5$
e. $y = x^3 + 3x^2 + x - 2$
f. $y = 6x - 14$
g. $y = x^2 + 2x - 5$
h. $y = 2x^3 - 4x + 1$

2. **TRIANGULAR NUMBERS**

$D = (1/2)n^2 + (1/2)n$; 3828 dots in 87th figure

3. **RECTANGULAR NUMBERS**

$D = n^2 + n$; 5402 dots in 73rd figure

4. **PENTAGONAL NUMBERS**

$D = (3/2)n^2 - (1/2)n$; 4510 dots in 55th figure

5. **DIAGONALS**

$D = (1/2)n^2 - (3/2)n$

6. **GREAT PYRAMID OF ORANGES**

22,100 oranges

CHAPTER 14, CLASSIC PROBLEM

8. **SUM OF CUBES**

25,502,500

Chapter 14, Problem Set A, Version 2

1. **A COVEY OF FUNCTIONS**

a. $y = 4x + 3$
b. $y = -5x + 4$
c. $y = 4x^2 - 2x + 3$
d. $y = (1/2)x + 7$
e. $y = -2x + 2$
f. $y = 2x^2 + 3x + 7$
g. $y = 2x^3 + 3x^2 + 7x - 1$
h. $y = (-3/2)x - 2$

2. **BLOCK PYRAMID**

2381 blocks

3. **A BEVY OF FUNCTIONS**

a. $y = 2x + 1$
b. $y = -x^2 + 4x$
c. $y = 4x - 7$
d. $y = x^2 + 4x - 3$
e. $y = -3x + 1$
f. $y = (1/2)x^2 - 6x - 2$

4. **HEXAGONAL NUMBERS**

$y = 3x^2 - 3x + 1$

5. **A GAGGLE OF CUBIC FUNCTIONS**
   a. $y = -2x^3 + 3x^2 + 6x + 23$
   b. $y = -x^3 + 4x^2 + 8$
   c. $y = x^3 + 4x^2 - 3x - 5$
   d. $y = (1/2)x^3 - 6x^2 - 2x + 5$

Chapter 14, Problem Set B

1. **CELEBRATION TIME**
   $46.00

2. **WILSHIRE BOULEVARD**
   A–27–B–15–C–21–D–48–E (A = Ardith, B = Burris, and so on. The numbers represent blocks between houses.)

3. **FIVES AND ONES**
   Carol: $8; Monica: $6; Tomás: $5; Andy: $4; Jake: $2

4. **REGIONS IN A CIRCLE**
   5051 regions

5. **PRICEY PETS**
   The sign above the cat cage was wrong. A cat costs $27.75. (Fish cost $8, and dogs cost $50.)

Chapter 14, Problem Set B, Version 2

1. **WORMS**
   144 cases (actually 143.4), or 143 cases with 144 worms left over

2. **PRODUCE**
   5 apples

3. **ANTIFREEZE**
   17.3 quarts

4. **STRANGE NUMBER**
   85,714

5. **TWO SEQUENCES**
   Tied on 54th term; passed on 55th

Chapter 14, Problem Set B, Version 3

1. **OLD CHEVY**
   228.48 miles

2. **COUNTY FAIR**
   They spent $120.20 if you count the $4 they gave the children and $116.20 if you don't.

3. **HOW MANY SEGMENTS?**
   $1/2\, n^2 - 1/2\, n$

4. **MYSTERY SUM**

							9
					7	9	8
				8	6	8	7
		7	2	5	7	5	7	6
	4	9	6	3	6	6	4	6	5
8	5	8	5	4	7	5	3	5	4
+ 7	6	7	4	5	8	4	2	4	3
1 6	7	6	3	6	9	3	1	3	2

5. **HOLLYWOOD SQUARES**
Top row: Gregorio Wilson, Della Jefferson, Howard Taft
Middle row: Isabel Kennedy, Betty Nixon, Clara Madison
Bottom row: Ellen Pierce, Frank Lincoln, Anh Roosevelt

Chapter 14, Problem Set B, Version 4

1. **ANTS**
25 inches long

2. **SEQUENCE RACING**
Passed on 221st term

3. **CORNER CAFÉ**
Clockwise around the table: Gil, french toast, sausage; Stan, omelette, hash browns; Karen, eggs, muffin; Donna, pancakes, strawberries

4. **THE GREAT NUMBERINI'S NUMBER**
38,095

5. **SKIING ARIZONA**
11 vehicles; 11 vehicles

Chapter 15, Problem Set A

1. **COFFEE STAIN**
Adults: $13; children: $7. The 58 should be 53, the 57 should be 54, and the 110 should be 118.

2. **THE OTHER THREE SQUARES**
33, 18, and 24

3. **TWO BILLS**
40, 41, 42, or 43 years old

4. **FAIR AIRFARE**
$198, $204, or $213

5. **AHSME**
Under both systems, guess on any question you can narrow down to two choices. Under the old system, guess on any question you can narrow down to three or four choices.

6. **TWO-INPUT FUNCTION**
$4x + y^2$

7. **SIXTEEN-TEAM TOURNAMENT**

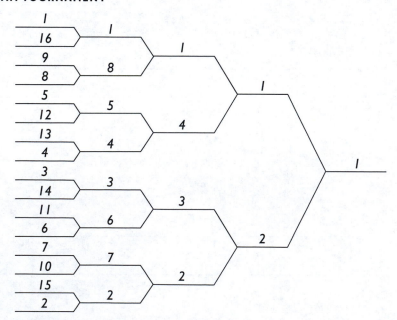

8. FOURTEEN-TEAM TOURNAMENT

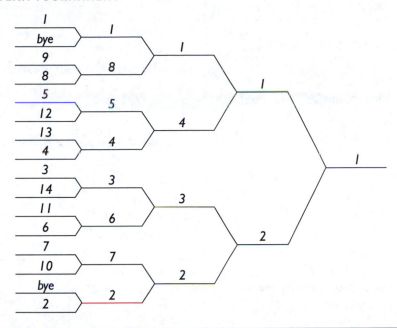

Student tree diagrams for problems 9–18 will vary.

9. ORDERING PIZZA

36 pizzas

10. INTERVIEW ORDER

24 orders

11. BAG OF MARBLES

a. White, $5/10$; not red, $8/10$

b. Both are white, $2/9$; one is red and the other is blue, $2/15$

c. $11/60$

12. BUSY SIGNAL

$20/32$

13. LAWN MOWER

He should not accept the deal. His expected value for drawing two bills out of the bag is $12. His normal rate of pay is $13.

14. FOUR ACES

$10/12$

15. X RAY

$96/100$

16. WORLD SERIES

$11/16$

17. SONS AND DAUGHTERS

$2/3$

18. DISEASE

.679

CHAPTER 15, CLASSIC PROBLEMS

19. NEW STATIONS

2 new stations. There were originally 11 stations.

20. THE MONTY HALL PROBLEM

You should switch. If you switch, you will win $2/3$ of the time. If you don't switch, you will win $1/3$ of the time.

Chapter 15, Problem Set A, Version 2

1. **THE CLASSIC HOMEWORK EXCUSE**
   Donut, $0.45; coffee, $0.65.
   5 coffees and 7 donuts should be $6.40.
   6 coffees and 11 donuts should be $8.85.
   9 coffees and 5 donuts should be $8.10.
   10 coffees and 8 donuts should be $10.10.

2. **SON OF TWO-INPUT FUNCTIONS, PART 1**
   a. $2x + y$
   b. $3y - x$
   c. $y - x$
   d. $2x - y$

3. **SON OF TWO-INPUT FUNCTIONS, PART 2**
   a. $2x + 3y$
   b. $x^2 + y$

4. **DOLORES'S AGE**
   11 to 13 years old

5. **LOG RIDE**
   Youngest, 14; oldest, 33

6. **COMPUTER PRINTOUT**
   Child, $4; adult, $6

7. **FREEZER**
   a. $20/156$
   b. $80/156$
   c. $136/156$

8. **CHIPS**
   a. $15/56$
   b. $6/56$
   c. $30/56$
   If the second chip drawn is blue, the probability that the first chip drawn was red is $15/35$.

9. **BASKETBALL AND MATHLETES**
   a. $8/35$
   b. $3/7$

10. **PAT PLAYS**
    $15/20$

11. **GLASSES**
    $11/29$, or about 38%

12. **TARGET SHOOTING**
    $4/13$, or about 31%

13. **MYXEDEMA**
    $90/1589$, or about 5.7%

14. **STAMP COMBINATION**
    3, 7, 8, 10, 11, 13, 15, 17, 18, 20, 21, 25, and 28 cents

15. **GOOD AND YUMMY RESTAURANT**
    a.

b.

	Jack cheese	Cheddar cheese	Hot sauce	Barbecue sauce
	X			
		X		
	X	X		
	X		X	
		X	X	
	X	X	X	
	X			X
		X		X
	X	X		X

Chapter 15, Problem Set B

1. **CALCULUS AND FRENCH**
   30 students; 20 students

2. **MOVIE THEATER**
   35 moves

3. **LOTSA FACTORS**
   1050 factors

4. **DICEY DIFFERENCES**
   1 occurs in 10 ways.

5. **AREA AND PERIMETER**
   5 by 20; 6 by 12; 8 by 8

Chapter 15, Problem Set B, Version 2

1. **ALONA'S BOARD FENCE**
   162.5 boards

2. **STOCKS**
   Cayla, $15; Billie Jo, $19; Alex, $57;
   Derwood, $60

3. **ZNORS**
   44 Zmuds. 8 Zmuds are also Zlogs and
   Zorfs.

4. **CONFERENCE**
   From Topeka to Peoria

5. **FOUR BY FOUR**

7	16	13	5
15	9	2	12
4	1	10	14
11	6	8	3

Chapter 15, Problem Set B, Version 3

1. **DEAR OLD DAD**
   Uncle Pedro is 24. Aunt Jasmine is 37.
   (Dad is 36, and Grandma is 58.)

2. **SIXTEEN CANDLES**
   PBPBPBPBGYGYGYGY or reverse

3. **REAL DOLLS**
   14 male, non–Cabbage Patch dolls

4. **WHH**
   Pairs, in order (h, w): (0, 0) (1, 2) (1, −1)
   (−9, −3) (−8, −4) (−9, −6)

5. **ONE HUNDRED FACTORS**
   45,360

Chapter 15, Problem Set B, Version 4

1. MARBLES

51 marbles

2. COUSINS

Many arrangements are possible.
Here is one:

Day 1

Hike	Gleane	Ricanna
1	Tom	Ryan
2	Bonnie	Torrey
3	Janet	Will
4	Stefan	Daniel
5	Michael	Gary

Day 2

Hike	Gleane	Ricanna
1	Michael	Torrey
2	Tom	Will
3	Bonnie	Daniel
4	Janet	Gary
5	Stefan	Ryan

Day 3

Hike	Gleane	Ricanna
1	Stefan	Will
2	Michael	Daniel
3	Tom	Gary
4	Bonnie	Ryan
5	Janet	Torrey

Day 4

Hike	Gleane	Ricanna
1	Janet	Daniel
2	Stefan	Gary
3	Michael	Ryan
4	Tom	Torrey
5	Bonnie	Will

Day 5

Hike	Gleane	Ricanna
1	Bonnie	Gary
2	Janet	Ryan
3	Stefan	Torrey
4	Michael	Will
5	Tom	Daniel

3. ANT GRAPEVINE

After 30 days

4. LONG LIST OF NUMBERS

Two thousand two hundred two

5. ALPHABET SOUP

Shape (or *phase*)

Chapter 16, Problem Set A

1. MORE DOTS

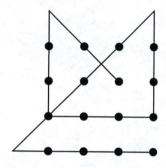

2. FEARLESS FLY

12.5 miles (change the focus to how long the fly is flying)

3. TOOTHPICKS

A tetrahedron will form four equilateral triangles. (A square with diagonals is acceptable if you recognize that the diagonals are longer than the sides.)

4. PERFECT SQUARES

97 perfect squares

5. COMPLEMENTARY EVENTS

a. Roll a die one time and don't get a 6.

b. Roll a die one time and get an even number.

c. Roll a die twice and get at least one 6.

d. Roll a die five times and don't get a 6 every time.

e. There are 40 people in the same room and at least 2 have the same birthday.

6. PAYDAY

$222.50

7. THE LIKELIHOOD OF BEING LATE

6.9%

8. ANOTHER CARD ARRANGEMENT

In order, from top to bottom: 3, 8, 7, A, Q, 6, 4, 2, J, K, 10, 9, 5

9. KNIGHT MOVES

a. 16 moves

b. Impossible

10. NEW GAME IN CENTRAL PARK

HOT	TANK	TIED
FORM	HEAR	BRIM
WOES	WASP	SHIP

Rotations and reflections are acceptable.

CHAPTER 16, CLASSIC PROBLEMS

11. WOULD HE HAVE SAVED TIME?

No. The time going half the distance by ox team is exactly the same as the time walking the entire distance. Even though the train is faster, its speed is more than completely offset by the slowness of the ox team.

12. THE CAMEL PROBLEM

533⅓ bananas or 534 bananas if the camel eats a banana before walking a mile

13. MARTINI GLASS

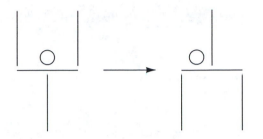

Chapter 16, Problem Set A, Version 2

1. BASIC HEX

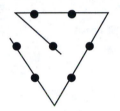

2. TEN-DOT PUZZLE

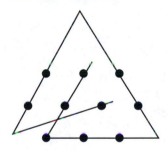

3. FOURTEEN-DOT PUZZLE

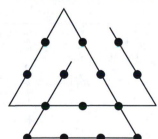

4. HOURGLASS

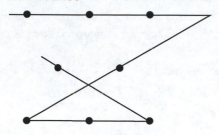

5. FRUIT STAND
$0.44 per pound

6. BADMINTON TOURNAMENT
28 ways

7. DANCING HEARTS
Only 1 group

8. AREAS OF SHADED REGIONS
$32 - 4\pi$, or approximately 19.4 square meters

9. MORE AREA
$6.25\pi - 6$, or approximately 13.6 square inches; yes, 5 inches is the length of the diameter.

10. ATTENDANCE FIGURES
630 student-days

11. LIGHT BEARINGS
5 weighings

12. HEAVY BEARING
3 weighings

13. PADUCAH HALF-TIME
Gabby, Izzy, Aracel, Evan, Kara, Britta, Henny, Fran, Charissa, Les, Johanna, and Darryl

14. WALKING HOME FROM SCHOOL
13.5 miles

15. MORE TOOTHPICKS

16. PERFECT CUBES
211 perfect cubes

17. SOME COMPLEMENTARY EVENTS
a. A family has three kids; at least one is a boy.
b. They watch three movies; two or fewer have "Friday the Thirteenth" in the title.
c. You find six coins, and at least one is a dime.
d. You meet four new people, and three or fewer are from Pennsylvania.

18. RUNNING GEORGE
26.5 miles

19. IS SHE OR ISN'T SHE?
63.2%

20. ALPHABET FLASH CARDS
I O A S J B Y P C K V D T L E Q Z F M X G R N H U W

Chapter 16, Problem Set B

1. COMPUTER ERROR
959 numbers

2. PALINDROME CREATOR
188 numbers

3. THE AMAZING RESTIN
STYLE

4. MULTIPLES

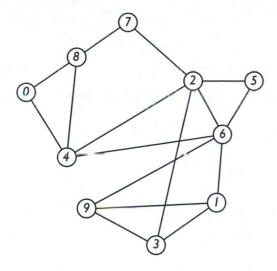

5. BOAT TRIP
15 weeks

Chapter 16, Problem Set B, Version 2

1. RECTANGULAR TILES
Many arrangements are possible. The large rectangle measures 23 × 31.

2. CONSECUTIVE INTEGERS
All powers of 2 less than 600: 1, 2, 4, 8, 16, 32, 64, 128, 256, and 512

3. ALPHABETICAL NUMBERS
Six hundred fifteen

4. SAY THE MAGIC WORD
WIND

5. WHERE SHOULD YOU STAND?
189th position

Chapter 16, Problem Set B, Version 3

1. LARGE CORPORATION
72.2%

2. MOVIE STARS
Bill Preston, red striped; Ted Reaper, green checkered; Don Logan, blue print; Grim Thanes, yellow plain

3. SUMSUMS
448 numbers

4. WRESTLING NEWSLETTER
LUNGE

5. THREE-DIGIT TRIANGLES

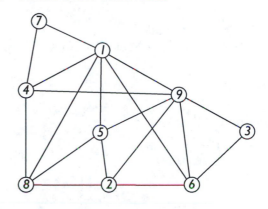

Chapter 16, Problem Set B, Version 4

1. REUNION
61 miles per hour

2. POKER
First hand: 6D, 6H, 6C, KD, KH

Second hand: 7, 8, 9, 10, J of diamonds (straight flush)

Third hand: 8H, 9H, 10H, JH, QD

Fourth hand: 2H, 2D, 3H, 3D, 4D

Fifth hand: 4, 5, 7, Q, A of hearts (flush)

3. BAND
5 × 12, or 6 × 8

4. DICE PRODUCTS
$7/11$

5. BOOK CHOICE
M

Chapter 17, Problem Set A-1

Student graphs for each problem will vary.

1. **REFRIGERATOR**
   The Major brand would be cheaper after 20 months. Because refrigerators are expected to last far longer than that, the Major brand is the most cost-effective. (She may also want to look at repair records for the two brands.)

2. **SODA**
   Restaurant: about $0.90; mini-mart: about $0.84

3. **LETTUCE**
   About 4 days

4. **APPLE ORCHARD**
   50 workers: 2.4 hours; 6 workers: 20 hours; around 9 to 12 people (answers will vary)

5. **JEANNE'S ORIGAMI BOOK**
   44 books

6. **CHRISTMAS TREE LOT**
   50 feet by 100 feet gives the maximum area of 5000 square feet.

7. **BOX**
   About 66 in.^{3}

8. **JAWS**
   About $5 million

9. **MORE PHONE CALLS**
   Day: $0.42, $0.33; evening: $0.30, $0.26; night: $0.22, $0.18

10. **BIG PROBLEMS**
    About 20 weeks

11. **SAILING**
    About 55 minutes

Chapter 17, Problem Set A-2

Student scale drawings for each problem will vary.

12. **YOUR BEDROOM**
    Answers will vary.

13. **TELEPHONE POLE**
    28 feet

14. **STADIUM POLE**
    73 feet

15. **HOW WIDE IS THE RIVER?**
    70 feet, or 23.3 yards

16. **FRISBEE ON THE ROOF**
    30 feet from the building, 25 feet off the ground

17. **KITE STRING**
    97 feet

CHAPTER 17, CLASSIC PROBLEM

18. **SAID THE SPIDER TO THE FLY**
    40 feet. Assume the north and south walls are the ones that are 30 feet long. Unfold and flatten the box, which is the room, so that the bottom of the west wall is connected to the floor, the floor is connected to the south wall, the south wall is connected to the ceiling, and the ceiling is connected to the top of the east wall. Then draw a line from the spider to the fly. This line will be the hypotenuse of a right triangle with legs 24 (6 across the west wall + 12 up the south wall + 6 across the east wall) and 32 (1 to the floor, 30 across the south wall, and 1 down the east wall to the fly). The distance is 40 feet.

Chapter 17, Problem Set A, Version 2

1. **DENISE'S AUTO REPAIR**
   a. 12-hour return, $120/hour
   b. 60-hour return, $24/hour
   c. 72-hour return, $20/hour

2. **RUNNING ON M. T.**
   a. 198 minutes
   b. About 8.5 miles

3. **KATHY'S CATERING SERVICE**
   Answers will vary. About $14.00 each for 10–19 people, about $13.00 each for 20–29 people, and about $12.50 each for 30–39 people.

4. **GROUND TURKEY**
   Answers will vary.
   a. 15 hours
   b. 6 months
   c. 47°

5. **SELLING SODAS AT THE PARADE**
   2 six-packs

6. **PIZZA**
   a. Yes, the graph is increasing and gradually turning downward or flattening out.
   b. Yes, the graph is turning downward or flattening out at a faster rate than the price-versus-diameter graph.
   c. About $15.50 using the diameter graph; about $16.00 using the area graph. Answers will vary for the real price charged.

7. **THE CHICKEN COOP**
   120 feet by 60 feet

8. **LAUNCHING ROCKETS**
   26 feet

9. **ROCKET TO THE MOON**
   360 feet

10. **DELBERT'S LAUNCH**
    40 yards

11. **TRAVERSING THE TRIBUTARY WITH CANINES**
    47 yards

12. **FLYING PAPER AIRPLANES**
    114 feet

13. **JANET'S PAPER AIRPLANE**
    282 feet

Chapter 17, Problem Set B

1. **SODA JERK**
   21; 52%

2. **LICENSE PLATES**
   3024 (the 3025th will have all different letters and digits)

3. **KAYAKING**
   $3^1/_4$ miles

4. **THE DIGITAL CLOCK AND THE MIRROR**
   22 times per day

5. **THE LATTICE**
   138 feet

Chapter 17, Problem Set B, Version 2

1. **BOXCARS**
   1. Back up right, unhitch gondola and tanker, and drive forward.
   2. Back up left, unhitch flatcar, drive forward.
   3. Back up right, pick up gondola, drive forward.
   4. Back up left, unhitch gondola and boxcar, drive forward.
   5. Back up right, pick up tanker, drive forward.
   6. Back up left, pick up boxcar, gondola, and flatcar, and he's done.

2. **THEY'RE TWINS**

Lynn: 23 three-point field goals, 3 two-point field goals, 9 free throws; Laurie: 5 three-point field goals, 25 two-point field goals, 19 free throws

3. **RUDY'S ROOT BEER**

21 days

4. **DIGGING A TUNNEL**

In $10^{1}/_{9}$ days

5. **TOASTER SERIAL NUMBER**

3224 toasters (the 3225th will have all different letters and digits again)

Chapter 17, Problem Set B, Version 3

1. **BEDTIME**

306 times per day

2. **CLASSIC CAR CLUB**

Red sports car

3. **BACK IN SHAPE**

564 loads

4. **THE WEDNESDAY CLUB**

13 members

5. **WOLF MOUNTAIN SKIER'S PLAN**

8 weekdays, 7 weekend days

Chapter 17, Problem Set B, Version 4

1. **HUESOS DE GALLO**

LAUGH

2. **WOLFIES ON THE RUN**

18 games

3. **WHEAT FIELD**

6.8 miles

4. **DIGITAL CLOCK**

124 times

5. **PERFECT SHUFFLE**

8 perfect shuffles

Index of Problem Titles

Page references in *italics* refer to a problem's solution.

Index

Digit-Place Game, 19–21
disjoint relationships, 216
Divisors and Reciprocals problem, note, 163
The Dog, the Goose, and the Corn problem, note, 176
Down on the Farm problem
    note, 71
    questions from the reading, 74, 76
Downtown Deli problem
    note, 71
    questions from the reading, 74, 76
A Drive to the Lake problem, note, 188–189
Drug Testing problem, note, 262

E

easier related problems (ERP), 161–173
    computers and, 163
    discussion, 161–162
    key question to ask, 162
Eggs in a Basket problem, note, 72
eliminating possibilities, 67–90
    conundrums, 67–70
    matrix logic and, 91
    seeking contradictions, 67
English-to-metric conversions, 146
equations
    finite differences and, 248
    guess-and-check and, 234–235
estimation
    easier related problems and, 162
    matrix logic and, 91
evaluation. *See* assessment

F

factors, 280
Farmer Ben problem
    note, 47
    questions from the reading, 50, 51
fascinating facts, group-building with, 13
final exam
    discussion, 319–320
    percentage of course grade, 40, 319
    problem options, 321–349
Finished Product problem, note, 59
finite differences, 248–259
    discussion, 248
focus. *See* Changing Focus, as theme
Ford, Henry, 23
formats for course, 41
Four Contiguous Stamps problem, note, 177
fractions
    guess-and-check and, 234
    unit analysis and, 145, 146
    working backwards and, 203–204
Franklin, Benjamin, 23
Frisbin problem, questions from the reading, 61
functions, finite differences and, 248

G

goals and philosophy of course, 2–3, 24, 38
Golf Match problem, scoring of, 31–37
Good Luck Goats problem, note, 163
Goodell, Suzanne, 81, 82
grading. *See* assessment
graphs, 294–295
group-building activities
    Color Square Game, 14–19
    Digit-Place Game, 19–21
    fascinating facts, 13
    interview, 12
    jigsaw puzzle, 12
    making 24 (or other number), 13
    name test, 13
    schedule of course and, 41
    the 1776 Game (or other year), 13
    teamwork (word game), 13
groups
    absent students and, 11
    benefits of, 9–10, 38
    changing of, 11–12
    Color Square Game in, 18
    desk arrangement for, 10–11
    Digit-Place Game versions for, 21
    final exam given to, 319–320
    instructor's role and, 9, 12
    as new approach, 9
    odd number of students and, 11
    percentage of grade for, 40
    Problem Sets B solutions within, 25–26, 42, 43
    safe atmosphere for, 5
    selection of students for, 11
    See also group-building activities
guess-and-check, 118–131
    algebra and, 234–235
    bracketing, 119
    charts for, 118
    discussion, 118–119
    fractional answers and, 234
    patterns and, 118
    physical representations and, 174
    quiz directions for, 120
    resistance of students to, 119

H

Haywire problem, note, 48
Holiday Blocks problem, note, 186
homework
    Problem Sets B as, 4
    Questions from the Reading and, 48
Hoop Greeting problem, note, 176
Horse Trader problem, note, 176
The Hotel Bill problem, note, 176
How Many Squares? problem, note, 162

patterns, 105–117
    discussion, 105
    guess-and-check and, 118
    systematic lists and, 58
Paying the Bills problem, note, 92
peers and problem solving, 9
Penny's Dimes, Part 1
    note, 59
    questions from the reading, 61, 62
Penny's Dimes, Part 2, note, 71
Persis's Gift Shop problem, note, 176
philosophy and goals of course, 2–3, 24, 38
Phone Calls problem, note, 294, 295
physical representations, 174–202
    Act-It-Out Day, 176
    changing of, 278
    conversion of problems into, 183–190
    discussion, 174–175
    using manipulatives, 177
    write-ups for, 175, 177
Pierce, Erin, 69
Pizza problem, note, 295
point of view, changing, 278–279
polynomial functions, 248
The Pool Deck problem, questions from the reading, 49, 51
presentations by students
    assessment of, 8
    on the boards, 6
    format of, 6
    frequency of, in classroom, 4, 43
    frequency of, per student, 8
    instructor behavior during, 6–7
    on overhead projector, 6
    percentage of course grade, 40
    time length of, 6
Problem of the Day (POD)
    assessment of, 6, 8, 40
    percentage of course grade, 40
    presentation of solutions. See presentations by students
    schedule of course and, 41–42
    sources for, 6, 7
    time required for, 6
Problem Sets A
    as homework, 4
    percentage of course grade, 40
    use of, 3
Problem Sets B
    alternates, 4
    assessment of. See assessment of Problem Sets B
    group solution of, 25–26, 42, 43
    as homework, 4
    prior preparation requirement, 43
    time in class for, 42, 43
    use of, 4
Problem Vats problem, note, 178
Problem-Solving Experiences in Mathematics (Charles), 26

proof by contradiction. *See* seeking contradictions
public policy, 147
Punk Blocks problem, note, 186
"A Puzzle for Pirates" (Stewart), 203

Q

quadratic functions, 248
Questions from the Reading
    Chapter 1, 49–51
    Chapter 2, 61–62
    Chapter 3, 74–76
    use of, 48
questions, student fear of asking, 9
quizzes, 38–39
    calculators allowed in, 39
    percentage of course grade, 40
    philosophy of, 38
    scoring method recommended for, 38
    source material for, 38–39
    time length of, 42
    timing of, in schedule, 38, 42
    See also final exam; midterm exam

R

random flailing, 58
ratios, units in, 145
reasoning, importance of, 24
representation, physical. *See* physical representations
resistance of students
    easy problems missed due to, 47
    guess-and-check and, 119
    manipulatives and, 175
    systematic lists and, 58
    written work documenting, 47
"The Riddle of the Vanishing Camel" (Stewart), 163

S

safe atmosphere, creation of, 5, 7, 8
Sallee, Tom, 19, 119, 161
Saturday at the Five-and-Dime Garage Sale problem, note, 235
scale drawings, 294
schedule of course, 41–43
    for 3-unit course, 42, 43
Scientific American, 163, 203
scoring. *See* assessment
seeking contradictions (indirect proof), 19, 67, 91, 262
1776 game (or other year), group-building with, 13
Simpletown Elections problem, note, 162
Soccer Game problem, note, 7
solution write-ups. *See* write-ups of solutions
some (as term), 216
Spatial Organization, as theme, 46, 175, 216, 294–310
spatial visualization
    diagrams and, 46
    physical representations and, 175